Volume 3

# APPLIED GEOMORPHOLOGY

# APPLIED GEOMORPHOLOGY
## Binghamton Geomorphology Symposium 11

Edited by
RICHARD G. CRAIG AND JESSE L. CRAFT

Routledge
Taylor & Francis Group
LONDON AND NEW YORK

First published in 1982 by George Allen & Unwin Ltd

This edition first published in 2020
by Routledge
2 Park Square, Milton Park, Abingdon, Oxon OX14 4RN

and by Routledge
52 Vanderbilt Avenue, New York, NY 10017

*Routledge is an imprint of the Taylor & Francis Group, an informa business*

© 1982 R.G. Craig, J.L. Craft and the contributors

*British Library Cataloguing in Publication Data*
A catalogue record for this book is available from the British Library

ISBN: 978-0-367-18559-6 (Set)
ISBN: 978-0-429-19681-2 (Set) (ebk)
ISBN: 978-0-367-46178-2 (Volume 3) (hbk)
ISBN: 978-0-367-46189-8 (Volume 3) (pbk)
ISBN: 978-1-00-302746-1 (Volume 3) (ebk)

**Publisher's Note**
The publisher has gone to great lengths to ensure the quality of this reprint but points out that some imperfections in the original copies may be apparent.

**Disclaimer**
The publisher has made every effort to trace copyright holders and would welcome correspondence from those they have been unable to trace.

# APPLIED GEOMORPHOLOGY

EDITED BY: **RICHARD G. CRAIG**
*Kent State University*

**JESSE L. CRAFT**
*U.S. Bureau of Mines*

*A proceedings volume of the eleventh annual "Binghamton" geomorphology symposium held at Kent State University, Ohio, October 2-5, 1980*

**GEORGE ALLEN & UNWIN**
London • Boston • Sydney

First published in 1982

GEORGE ALLEN & UNWIN LTD
40 Museum Street, London WC1A 1LU

**British Library Cataloguing in Publication Data**

Applied geomorphology.—(The "Binghamton"
 symposia series in geomorphology. International
 series; no. 11)
 1. Geomorphology—Congresses
 I. Craig, Richard G.   II. Craft, Jesse L.
 III. Series
 551.4   GB400.2

ISBN 0-04-551050-4

**Library of Congress Cataloging in Publication Data**

"Binghamton" Geomorphology Symposium (11th: 1980:
 Kent State University)
 Applied geomorphology.
 (The "Binghamton" symposia; 11)
 Includes index.
 1. Geomorphology—Congresses. I. Craig, Richard G.
 II. Craft, Jesse L.   III. Title.   IV. Series.
 GB400.2.B56 1980    551.4    81-22791
 ISBN 0-04-551050-4           AACR2

Printed and bound in Great Britain by
Mackays of Chatham Ltd

# PREFACE

Geomorphologists tend to consider all geomorphologic research to be "applied." In the sense that each advance in knowledge provides a clearer view of how the earth works, we must all be applied scientists. Yet too much of our work remains obscure and poorly understood by those who are most likely to benefit from it. Very important research remains unavailable "theory" only because it is difficult to interpret and many of us are too busy extending our own research to help others make use of it. Thus the purpose of this book is not to codify a distinct set of methods and data that represent a branch of knowledge called "applied geomorphology." Rather it is designed to show geomorphology as it is (and can be) applied to current problems facing the people of the world.

We have tried to make this sampling a representative one. But that means weighting the possible contributions; and weights can be made in several ways. We chose to emphasize problems that are encountered where man lives on the one hand and where man interacts with the surface through important construction projects on the other.

A large part of the world's population lives near the ocean. So processes shaping that environment must be of prime consideration for many of us. Accordingly, four chapters are devoted to various coastline phenomena. It is important to recognize that shore erosion is not the only problem facing us. Coastal subsidence can also have devastating effects. Thus our sampler covers erosion along oceans where emergence is occurring (Chapter 11), where submergence is occurring (Chapter 9), along large inland bodies of water where the processes have a special mix (Chapter 10), and finally we look at the case of subsidence accompanied by deposition (Chapter 12).

Another morphogenetic region with which man is constantly interacting is that area glaciated only recently. Here a number of problems exist, including: where to live, how to construct facilities where climate is extreme and glacial and periglacial processes still active, and how to dispose of our wastes. Fiksdal outlines methods to deal with the first (Chapter 3), Clague (Chapter 2) presents an excellent example of the second, and Hatheway and Bliss (Chapter 4) deal with the last.

Two other environments of immediate concern to man are more restricted but present uniquely exasperating problems. Carbonate terrains, which may have been avoided in the past, are now increasingly subject to development. Sheedy and coauthors (Chapter 13) describe some situations that the applied geomorphologist encounters in such terrain. A second area of growing importance is presented by the problems unique to certain areas in the midwest and south-central parts of the United States. The particular problem of expanding soils has grown acute because of the building boom occurring in states such as Texas. The lesson to be learned here is that the job of the geomorphologist requires that (s)he be both specialized and adaptable. As societies needs are redefined, so must be geomorphologists' abilities.

We include a set of examples of specific applications of geomorphic knowledge and techniques. A number of these result from the increasing scale of construction projects: for example, the geotechnical problems associated with the MX missile problem (Chapter 7) and the giant natural gas pipeline project (Chapter 2). A third example arises when we contemplate the disposal of radioactive waste in salt domes. A necessary condition is that these domes must be demonstrably stable on the time scale of tens of thousands of years. This question is considered in Chapter 5.

Computer applications in geomorphic studies are becoming increasingly important. Vanderpool (Chapter 14) shows how the computer can be used to study the stability of landforms. Craig (Chapter 8) considers some geomorphic constraints that must apply in such models.

Finally, we present two excellent examples of geomorphic thought applied to features of the fluvial system. Bernard and Melhorn (Chapter 15) show the long-term (on the human scale) response of streams to engineered "improvements." Stall and Herricks (Chapter 16) show how the natural dynamic variability in a stream (such as the pool and riffle sequences and changing flow regimes) exercises strong controls on the biota. This will undoubtedly fill a gap in our appreciation of the delicacy of this habitat.

This book makes another point. As you review the list of authors you will undoubtedly be struck by the preponderance of multiple-authored papers. Where a complex field of knowledge must be applied to an equally complex set of problems, the techniques frequently require the abilities of several practitioners. Meshing the relevant approaches is a team effort. We hope that this sampler improves your understanding of the importance of geomorphology in current problems and of the ways proper understanding of the geomorphic principles can benefit humankind.

RICHARD CRAIG
JESSE CRAFT

# CONTENTS

# CONTRIBUTORS

J. Aghassy *(deceased)*
Department of Geography, University of Pittsburgh

James C. Armstrong
Department of Civil Engineering, University of Texas at Arlington

Robert S. Barnard
Texas Oil and Gas Corporation, Corpus Christi, Texas

William F. Beers
Roy F. Weston, Inc., West Chester, Pennsylvania

D. J. Benson
Department of Natural Resources, Division of Geological Survey, Sandusky, Ohio (present address: Department of Geology and Geography, University of Alabama, University, Alabama)

Z. Berger
Law Engineering Testing Co., Marietta, Georgia (present address: Exxon Production Research Co., Houston, Texas)

Zenas F. Bliss
Haley & Aldrich, Inc., Consulting Geotechnical Engineers and Geologists, Cambridge, Massachusetts

C. H. Carter
Ohio Department of Natural Resources, Division of Geological Survey, Sandusky, Ohio

Gary E. Christenson
Fugro National, Inc., Long Beach, California; presently Utah Geological and Mineral Survey, Salt Lake City, Utah

John J. Clague
Geological Survey of Canada, Vancouver, British Columbia, Canada

William F. Cole
Engineering Geosciences Research Program, Department of Geology, Texas A & M University, College Station

Richard G. Craig
Department of Geology, Kent State University, Kent, Ohio

Robert Dolan
Department of Environmental Sciences, University of Virginia, Charlottesville

A. G. Everett
Everett and Associates, Rockville, Maryland

Allen J. Fiksdal
Department of Natural Resources, Division of Geology and Earth Resources, Olympia, Washington

**D. E. Guy, Jr.**
Ohio Department of Natural Resources, Division of Geological Survey, Sandusky, Ohio

**Allen W. Hatheway**
Haley & Aldrich, Inc., Consulting Geotechnical Engineers and Geologists, Cambridge, Massachusetts

**Bruce Hayden**
Department of Environmental Sciences, University of Virginia, Charlottesville

**Edwin E. Herricks**
Department of Civil Engineering, University of Illinois, Urbana

**James T. Kirkland**
Department of Geosciences, University of Missouri at Kansas City

**Walter M. Leis**
Roy F. Weston, Inc., West Chester, Pennsylvania

**Christopher C. Mathewson**
Engineering Geosciences Research Program, Department of Geology, Texas A & M University, College Station

**Paul May**
Department of Environmental Sciences, University of Virginia, Charlottesville

**Suzette K. May**
Department of Environmental Sciences, University of Virginia, Charlottesville

**Wilton N. Melhorn**
Department of Geosciences, Purdue University, West Lafayette, Indiana

**James R. Miller**
Fugro National, Inc., Long Beach, California

**Prinya Nutalaya**
Division of Geotechnical and Transportation Engineering, Asian Institute of Technology, Bangkok, Thailand

**Denise D. Pieratti**
Fugro National, Inc., Long Beach, California

**Jon L. Rau**
Department of Geological Sciences, The University of British Columbia, Vancouver, British Columbia, Canada

**Katherine A. Sheedy**
Roy F. Weston, Inc., West Chester, Pennsylvania

**John B. Stall**
Consulting Research Hydrologist, Urbana, Illinois

**Abraham Thomas**
Roy F. Weston, Inc., West Chester, Pennsylvania

**N. Luanne Vanderpool**
Department of Geological Sciences, University of Illinois, Chicago

# 1

# GEOMORPHIC PROCESS DATA NEEDS FOR ENVIRONMENTAL MANAGEMENT

*A.G. Everett*

## ABSTRACT

Societal demands leading to an increase in the intensity of land use, coupled with protection of the environment, are requiring increasing use of data from physical and chemical surface and near-surface processes involving the atmosphere, lithosphere, and hydrosphere. These interrelated processes are attracting increased attention from many disciplines but foremost among them should be geomorphology, for which the traditional focus has been such processes and their interrelationships.

A variety of land uses contribute to the need for process rate data: mineral and energy development; industrial, commercial, residential, and recreational development; and waste repositories, especially for the more toxic or polluting forms of wastes. To minimize the costs of maintenance during use, and, perhaps even more important, long-term maintenance over centuries, landforms of long-term stability should be designed and processes having slow rates of change should be utilized. Data for such design are not readily available in the scientific literature. Quantitative data on the short-term rates of processes under varying conditions are exceedingly scarce and are not commonly available. The more readily available estimates are for periods of thousands of years. Contrasting short-term rates for processes common to arid, temperate, and humid climates are virtually unknown. Despite several centuries of work describing landforms qualitatively, methods for rigorous quantitative analysis of process rates, including their variability, are not extensively available for application to monitoring or design problems in surface and shallow subsurface environments.

In this paper the need for such data is stressed, using various examples of slope stability and of geochemical and material transport problems associated with coal and uranium mining in arid regions of the western United States, with lignite and coal reclamation in more temperate regions such as central Texas and the eastern United States, and with recent problems associated with mining under humid tropical conditions in Papua New Guinea. Tropical environments may serve as excellent natural laboratories for certain processes that will occur under temperate and arid conditions as well as humid, but over a longer time period.

## INTRODUCTION

The national objectives of many countries for increased rates of production of metallic ores and fuels, such as coal, oil shale, and uranium, coupled with concerns about environmental degradation in general and the handling of potentially toxic wastes in particular, will increase the need for careful planning and engineering of spoils, tailings, and waste material deposits. The large volume of many of these deposits creates the need for low maintenance cost, physically and hydrologically stable, man-made landforms that erode, degrade, and leach at geologically slow rates. It is desirable that such landforms be able to accommodate new uses; many, however, may need to be designed for minimal erosion and leaching rates in order to control rates of release of potentially toxic materials into the biosphere, thus precluding land uses that would require restoration to original conditions. The geographic, climatologic, and geologic settings for these man-made landforms cover the entire spectrum of conditions on the earth's surface. Economics and feasibility will impose substantial constraints on design in both the emerging countries and those whose economies are well developed. Low-erosion-rate landforms commonly will be more expensive to construct than are those commonly used today for waste disposal, but long-term maintenance and monitoring costs, as well as possibilities of catastrophic failure, should be much reduced.

Geomorphology, having as its traditional focus the study of landforms and their processes of formation, has a major contribution to make by providing the basic design data and concepts for the creation of man-made landforms having long-term stability. At the present time, process rate data are scant for calculating stability conditions under conditions of longer-term climatic change as well as short-term meteorologic variability and the resultant physical and chemical alteration of waste pile materials.

The accepted standards for the design of stable slopes are those resulting from soils, foundation, and rock mechanics engineering tests developed under laboratory conditions. In common application, these factors do not take into consideration the variations in precipitation, permeability, soil moisture, and mineral weathering conditions that may be expected to occur during the time that waste materials are exposed to surficial processes. Empirical factors of safety commonly provide stability for a short period of time, essentially that prior to the dynamic change of geologic characteristics. The shortcoming of laboratory testing in slope design has not gone unnoticed in the soils engineering literature, as indicated by the comments of both Bishop (1966), concerning progressive failure of stiff clays, and Bjerrum (1966), on problems of slope stability resulting from weathering of clays.

Traditional engineering design factors are applied to residual slopes remaining after excavation of mines, quarries, and other varieties of cut slopes as well as to waste piles. Slopes receiving uncompacted overbank deposits can change their stability conditions both extensively and rapidly. In mining projects, the economics of removal of overburden frequently dictates that slopes be maintained in a state of slow erosion or slumping at greater-than-optimum stable slope angles. Debris from such unstable slopes is often accommodated in temporary storage on inactive benches or cleaned from slope toes at a cost less than that of stripping to create pit walls stable over a long term. After the cessation of operations, such slumping of slopes will partially fill the excavation by sapping walls and adjacent surfaces. Obviously, these processes have their own hazards both during and subsequent to the active mine operation, but they usually can be carefully managed while operations remain active and profitable. At the conclusion of operations, however, geologic factors for long-term stability are needed in order for the excavations to serve alternative land uses.

# PHYSICAL AND CHEMICAL CHANGES THAT CAN AFFECT WASTE PILE PERMEABILITY

The engineering literature on the stability of rock slopes and of unconsolidated materials is growing rapidly, but much of it has not yet accounted for the dynamics of processes that will, with time, modify the original internal physical structure intended to provide long-term stability. Seed (1974) and Kealy and Busch (1979) have discussed the need to consider physical dynamic loading conditions in initial designs. They have focused both upon ruptures and liquefaction that can result from earthquakes and from "strain-induced internal movement," citing the mine refuse dam failures at Buffalo Creek, West Virginia, in February 1972 as one familiar example of inadequate consideration. Preventive measures that they suggest include good compaction and drainage. The same factors were emphasized as well by Brawner (1979) in his review of metal mine tailings dam designs, in which he considered the problems of liquefaction of sand-size material and concluded that, with adequate density achieved by compaction, "the increased sheer strength and stability of compacted sands allows much steeper slopes and lower safety factors to be used." Over time, modification by surface and near-surface physical and chemical processes can modify initial permeability (and particle composition) in waste materials (hence pore water pressures in piles and slopes) changing an initially sound design to one leading to failure.

Regulations of government agencies commonly call for revegetation of spoil and waste piles. The revegetation processes disrupt the initial compacted fabric by root penetration, leading to channel formation by root decay, and by alteration of alumino-silicate and carbonate components in contact with the hydrogen-ion sheath around rootlets (Loughnan 1969 has cited pHs below 4 and possibly as low as 2 adjacent to such rootlets). Revegetation of spoils will progressively modify the pore fabric by steadily increasing the volume disturbed, by root channels reoriented, or compacted by root pressure. Analogous modification of permeability and resultant soil moisture flow in natural soils has been discussed by Baker (1978) and by de Vries and Chow (1978). It is to be expected in revegetated spoil spiles and on cut slopes as well.

As noted by de Vries and Chow (1978), agricultural soils tend toward substantial horizontal homogeneity as a result both of their geologic origin, commonly as alluvial, eolian, or lacustrine deposits, and of repeated cultivation. Groenewold and Rehm (1980) indicate that spoil piles, like the forest soils studied by de Vries and Chow, tend toward both lateral and vertical inhomogeneity. Groenewold and Rehm report that, over time, both increases and decreases in waste pile hydraulic conductivity can occur. They have related these changes to compaction and to enlargement of voids. Although they have noted the relationship in spoils of highly dispersive sodic materials and piping, which may not begin for up to 5 years, chemical precipitation or small particle formation and movement in the interstices of the waste pile also may contribute to permeability modification (hence to pore water pressure distribution). In addition to interporosity chemical precipitation, another factor that may contribute to the formation of fine particles for movement within the waste pile are small particles formed by the freezing of water-saturated humic soil components. Giesy and Briese (1978) have formed fine humic particles under laboratory conditions by freezing pond water. Upon thawing the solutions and after sonic dispersal, the pond water retained increased particle sizes. Soil mulches, or other soil treatments with humic materials, and the normal accumulation of humic material with continued successful vegetation of spoils could provide the necessary humic materials to soil solutions for the generation of humic particles under freezing conditions. These particles could then move in pore spores to places where they would block permeability, leading to localized increases in pore water pressures.

Physical modification of waste pile permeability and internal pore water pressure may be a significant factor in many cases of waste pile and cut slope failure. Such physical changes, which may also be related to chemical changes, may not appear rapidly. The appearance of the same phenomenon in two places will differ in time, based on the rate at which the changes are taking place in the two locations. For those changes that occur in aqueous solutions in freely draining waste piles, humid tropical areas offer the opportunity for examination of some processes that are much accelerated. Strakhov (1967) estimated that increased leaching rates in the humid tropics increase weathering rates from seven to fourteen times that of temperate climates and, when the effects of higher average temperatures and more abundant organic matter are added to the increased rainfall, the differential in weathering rates may reach twenty to forty times. Whatever rates may best characterize the difference, it is apparent that weathering and transport processes dependent upon rainfall are much more rapid in humid tropical climates than in temperate climates. Therefore, changes in hydraulic conductivity of waste piles that influence their stability in the tropics is relevant to long-term design questions under other climatic conditions where similar processes take place under much reduced rates.

Observations by Meynink (personal communication, 1979; also Meynink 1979) on permeability changes in mine waste piles at Bougainville Copper Limited's Panguna Mine, Papua New Guinea, indicate that under rainfall conditions of about 4 m per year, mineralogical alteration and rock particle breakdown can progress sufficiently within 1 to 3 years to begin to plug parts of what were initially permeable, freely draining waste dumps. My observations at the Panguna Mine suggest that such permeability modification may be a significant contributing factor in the formation of debris flows and slump failure along dump slope surfaces. Such modifications of the internal permeability would behave similarly to secondary permeability in lithified rocks in redirecting water flowing downward to a horizontal direction where it then moves laterally to intersect weathered materials on a slope surface, either saturating the mass and creating liquefaction or so modifying pore water pressures as to create mass failure of other types (Everett 1979). Chemical weathering of the feldspar-rich volcanic mudflow and diorite prophyry waste pile components at Panguna produces amorphous aluminosilicate gels, kaolinite and/or halloysite, and hydrated iron oxides as well as reduction in the size of rock fragments by disaggregation into the component mineral grains (Everett et al. 1980a, 1980b). The weathering process produces a wide variety of grain sizes of altering minerals, rock debris, and weathering products in the dumps. It is this combined variety of materials that physically alters the internal permeability, hence ultimate stability, of the dumps by producing changes in flow paths and pore water pressures.

Although intense, tropical conditions such as described from Papua New Guinea are clearly quite limited insofar as potential sites for man-made landforms from waste deposits are concerned, analogous mineralogical and geochemical transformations leading to physical modifications of internal waste pile hydrology can be expected to evolve in aluminosilicate mineral-bearing waste deposits elsewhere with time. Under attack from acidic solutions formed by oxidation (especially bacterially enhanced oxidation) of sulfides, or even from acidic precipitation, aluminosilicate materials will undergo both chemical and physical modifications in temperate and even arid climates. The alteration process can be accelerated when the waste materials are watered to induce vegetation growth.

Except for Meynink's observations at the Panguna Mine, no estimates of short-term weathering and decomposition rates for the high rainfall Papua New Guinea area are available. Groenewold and Rehm (1980) provide additional information on the mechanisms leading to hydraulic conductivity changes and piping in spoil piles at sites in North Dakota, as well as indicating the time of onset of changes that can lead subsequently to pile failure. They indicate that compaction of spoils begins immediately and is most extensive in the first 12 to 15 months, reaching 0·5 to 1 m at one site and 0·15 to 0·6 m in 6 months at another site. At the latter site, greater compaction took place in dozer-contoured materials placed during the winter; compaction of "summer-dozer and mid-winter scraper-contoured areas has been minimal" (less that 0·07 m). Compaction at a reclaimed strip mine in western Maryland resulted in subsidence of about 0·3 m in the first year. Compaction in the absence of waste pile failure or erosion usually results in acceptable restoration with return to alternative land uses. Such reclamation is more likely to occur in areas of low relief with rainfall adequate to support vegetation without irrigation (Fig. 1). The rates at which compaction proceeds and the variables (such as spoil composition, grain size grading, method of emplacement, and degree of original compaction) affecting those rates are needed increasingly in waste pile design. The rates at which processes, such as weathering, affect pile characteristics under different conditions of climate and spoil composition are also needed for design consideration.

**Figure 1** Lignite strip-mined land in the foreground has been reclaimed as pastureland. Active stripping is in progress in the background. Big Brown Steam Electric Generating Station, near Fairfield, Texas.

## CLIMATIC CHANGES THAT MAY AFFECT
## LANDFORM DESIGN

Climatic factors in landscape formation obviously go beyond the question of weathering. These factors have been discussed extensively in texts and papers for many decades, as has the matter of landform response to climatic change. Consideration of climatic change can assume a more significant role in future landform design because climatic research is beginning to provide information on changes over time periods on the order of decades and many of the landforms to be developed will require low-erosion-rate characteristics over centuries. Rates of denudation of consolidated materials have been estimated on a regional basis in a variety of studies, as discussed by Arnett (1979), but rates of slope retreat or erosion for individual landform units, other than for catastrophic events such as landslides, are very poorly known and rarely estimated.

For landforms composed of potentially hazardous materials or materials with a high pollution potential, such low rates of erosion are clearly desirable. Examples of such materials are radioactive uranium mill tailings, oil shale retort wastes, synthetic fuel process wastes, and coal washing wastes. Materials such as oil shale retort wastes may have unusual physical characteristics causing them to behave in unexpected ways when placed in typical types of waste piles. Figure 2 shows one type of oil shale retort waste originally emplaced in a test plot under 0·6 m of sterile soil cover that was subsequently vegetaged. The waste material is hydrophobic rather than hydrophilic in character. Upon receiving water by infiltration through the vegetated soil cover, the retort waste flowed out as a thick slurry or mudflow. Such wastes, possibly containing hazardous compounds such as benzo(a)pyrene (Schmidt-Collérus 1974), need containment so that they are not subject

**Figure 2** Oil shale retort waste flow that moved from beneath 0·6 m of sterile soil material on a test plot embankment under spring meltwater and rainfall conditions, western Colorado.

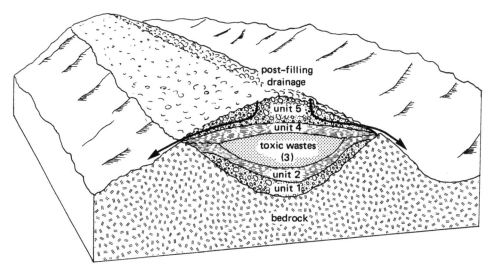

**Figure 3** Diagrammatic cross section of topographic reversal designed to contain toxic materials in a low-erosion-rate landform.

to erosion or contact with infiltration water. Everett et al. (1974) proposed a scheme for containment of such wastes, using the concept of an international topographic reversal, using an underdrain bed of gravel (unit 1) along the valley floors and walls, covered in turn by a well-compacted, very low permeability, clay layer underlying (unit 2) and surrounding (unit 4) the waste material (unit 3) (Fig. 3). Damming of the valley mouth in such a way as to allow drainage from the gravel drain but containing of the overlying clay blanket and waste material is required. To prevent vegetative disruption of the upper surface, a coarse gravel cap with large, freely draining void space (unit 5) must cover the clay blanket emplaced on top of the waste deposit in the core of the deposit. When the valley is filled and capped at a level above the flanking ridges, the coarse cap will resist erosion. A new erosion pattern, draining down the newly created ridge over the outer flanks of the enclosing valley walls, can be established. The previously adjacent valley walls can then handle runoff from the newly formed ridge, creating a new drainage and erosion system with the former valley now transformed into a drainage divide.

    Gravels resistant to erosion are not uncommon as cap rocks in drainage reversals that have occurred on the High Plains, where stream gravels have become established as remnant high terrain. Lattman (1977) has shown that caliche rubble, under arid conditions, also forms a resistant rock capping buttes. Man-made coarse gravel and boulder deposits have withstood erosion for about 100 years in such diverse climatic areas as the Mesabi Range (Fig. 4) and South Park, Colorado (Fig. 5). Both gravel and boulder deposits in these locations are well drained and unvegetated. A gravel cap has developed by erosion and deflation during the last 20 years on at least one uranium strip mine waste pile in the Gas Hills of Wyoming under arid conditions (Fig. 6). In none of these cases has vegetation reestablished; thus one source of permeability derangement and accelerated weathering has been prevented naturally. The lack of revegetation represents a reduction in carrying capacity for the local ecosystem but where ecologically deleterious wastes occur, an incremental decrease in carrying capacity is more desirable than marginal productivity associated with increased rate of availability of toxic components. The

**Figure 4**  Well-drained coarse gravel and boulder deposit of mine waste material, south-central Misabi Range, Minnesota.

**Figure 5**  Coarse gravel and boulder pile of gold-mining dredge spoil, South Park, Colorado.

**Figure 6** Gravel cap developed by deflation on a uranium strip mine waste pile, Gas Hills, Wyoming. Vegetation in the foreground is on undisturbed natural soil. Waste materials in the middle distance are largely sands with a gravel cap.

design of such landforms requires an understanding of both processes and rates to avoid competing processes from undermining the new landform. An example of the latter is gully erosion of the flanks of the uranium waste pile, undercutting the gravel pavement on the surface. Many waste piles in arid and semiarid environments are designed with slopes much steeper than those that typify natural landforms in the same environment. Sheetwash and rillwork, which rapidly develop into closely spaced ravines, form rapidly under arid and semiarid conditions, continued proof of the efficacy of water as the major erosional medium under arid conditions. Waste pile slopes that are unengineered (angle of repose) and revegetated naturally and those that are intentionally engineered and revegetated are shown in Figures 7 and 8, respectively. Both show substantial surficial erosion. In contrast with natural slopes in an arid intermontane basin region in southern Wyoming (Fig. 9), these man-made slopes are both steeper and more actively eroding. Although some angle-of-repose slopes in unconsolidated materials may show initial stability at quite high angles, established slopes with angles as low as 10 to 15% commonly show extensive surficial erosion over time in virtually all climatic conditions and states of revegetation.

Regevetation is clearly critical to the stability of fine- to medium-grained surface sediments but the year-to-year variability of natural vegetation density under arid to semi-arid conditions is so great that vegetation alone cannot prevent, or even substantially retard, surface erosion. Martin (1975) illustrates the annual variability that can be expected under differing elevation and rainfall conditions where annual grasses constitute a significant part of the herbage supply. Groenewold and Rehm (1980) observe that vegetation on reclaimed spoils in North Dakota under semi-arid conditions is typically thicker during the first year and shows a consistent decrease thereafter. Historically, this

**Figure 7** Waste pile slope showing extensive rillwork and gully erosion, near Tucson, Arizona.

**Figure 8** Waste pile slope with extensive gullying despite revegetation, northern Powder River Basin, Wyoming.

**Figure 9**  Profiles of natural colluvial slopes at the base of the southwestern side of the Kinney Rim, Wyoming.

has been a significant problem in a number of locations even with semiarid to temperate rainfall amounts. For that reason, vegetation cannot be counted on to stabilize slopes in the range 1 in 3 to 1 in 4, as commonly is now done on waste piles. Vegetation on strip mine spoils from coal mining operations in the western United States shows the ability to concentrate toxic elements such as mercury, lead, arsenic, selenium, fluorine, and molybdenum [Connor et al. 1976; Ebens & McNeal 1977; Erdman 1978). Should the potential for toxic concentrations of such elements in vegetation become a factor in land restoration decisions, revegetation may no longer be regarded as a sine qua non for surface restoration. Then, more than ever, long-term stability of man-made landforms under climatic variation must become a serious design consideration.

Climatic effects on landform stability are traditionally treated on the scale of major changes, such as those of glaciation. In the western United States, changes in aridity have been correlated with alluviation and gullying (Hack 1942; Bryan 1954). Recent studies of climatological change, such as that of Lamb (1977), provide a basis for correlation of recent climatological change with landform stability. Bradley (1976) provides an analysis of precipitation records from roughly 1850 through 1970 for the Rocky Mountain region and Lawson (1974) gives a methodology for reconstruction of the climate of the western interior of the United States from 1880 to 1950. Future data on both climatological variability and landform stability should be much easier to get, although local variability in rainfall intensity is quite high, especially in the mountainous western United States where much of the mineral development activity will take place. Based on his studies of climatic variation and directions of change, Reid Bryson (personal communication, 1974) has predicted that there will be increased local intensity of precipitation in the Rocky Mountain area for a period of years. Despite regional trends in climate, specific event data for mountainous areas during high-intensity rainfall periods is sparse, as shown by recent reconstructions of the storm associated with the flood of July 31–August 1, 1976, in the Big Thompson River and Cache la Poudre River Basins in Colorado (Simons et al. 1978; McCain et al. 1979). For the near future, average event intensity data can be used to correlate the stability of man-made landforms with climatic effects.

## EFFECTS OF MAN ON SEDIMENT LOADS AND RIVER MORPHOLOGY

Sediment transport models have flourished in the last three decades, stimulated in large part by the works of Vanoni (1946) and Einstein (1950; see Shen 1979 for a recent review of the state of the art of modeling). Rates of events under natural stream conditions are still difficult to predict with accuracy despite progress in quantitative prediction of magnitudes of transport and of deposition. Predictive modeling of morphological modifications of streams and their deposits is still in its early stages. From the viewpoint of the geomorphologist, there is great incentive for development of such quantitative predictive modeling. To turn briefly to Papua New Guinea, recent mining operations on Bougainville have given rise to some rapid and impressive modifications of the Kawerong and Jaba Rivers. The general problem of disposal of overburden and mill tailings, amounting initially to $10^7$ tonnes of sediment, has been described by Higgins (1977) and Pickup and Higgins (1979). In the first 7 years of sediment discharge, an estimated 65% of the material passed down the Kawerong River into the lower part of the Jaba River, then was deposited as a delta along the shore of Empress Augusta Bay. About 35% of the material was deposited in the Kawerong Valley and on the lower Jaba floodplain. To decrease the volume of sediment entering the system, the sediment discharge was modified in 1977 so as to discharge only tailings with the overburden previously dumped in the stream and eroded from tailing dumps now being left along the valley walls in dumps upstream of the tailings discharge. The reduction in load caused downcutting, so that in 2 years the Kawerong cut down as much as 10 m through the deposited waste material in the upper part of the Kawerong gorge, just downstream from the mill tailings discharge. The depth of downcutting toward the mouth of the Kawerong decreased to 2·5 m. Spectacular terraces were left in the inner gorge of the Kawerong (Fig. 10).

Man has frequently caused increases of sediment load in the course of land use, changing the rate of sedimentation and occasionally changing the morphological patterns of streams as well (Park 1977). Thornes (1977) observed, and we agree based on our attempts to quantitatively predict the onset and scale of morphological modifications to the Fly River system as a result of mining (Everett et al. 1980b), that "most hydraulic geometry work to date...[i]s rather empirical." However, with increased intensive land use, quantitative estimates of the magnitude of sedimentation, erosion, and morphological change in steams will be of increasing importance.

## CONCLUSIONS

The role of the geomorphologist as an interpreter of surface and near-surface phenomena affecting the earth's surface is frequently regarded as being largely of academic interest in the historical study of the Earth. As those involved in this symposium are aware, however, there are numerous potential applications of geomorphology – as the study of

**Figure 10** Terraces of mine waste rock, mill tailings, and river sediment along the inner gorge of the Kawerong River, Bougainville, Papua New Guinea, as a result of reduction in sediment load from the mining operation. The depth of downcutting in this view, on the order of 7 to 8 m, was accomplished in 2 years.

*processes* – to environmental problems. In this paper my focus has been upon surficial processes. I have used hypotheses, data, and conclusions from geochemistry, petrology, hydrology, and geomorphology in discussing surficial problems. There are those who prefer to treat geomorphology as a macroscopic system; they may be dismayed at the emphasis on microscopic and chemical phenomena in the course of unraveling geomorphic processes. Indeed, Chorley (1978) has expressed dismay over the extent of the present emphasis in geomorphology on such utilitarian, short-term studies as those which are associated with environmental management. Brunsden et al. (1978) provide a more optimistic view of the utility of applied geomorphology in Britain, citing a variety of problems to which geomorphic hypotheses, data, and concepts have important application. As the quantitative treatment of processes progresses in assessment of rates and their variability, and in evaluating concurrent or competing processes in terms of relative intensity or effectiveness per unit of time, traditional megascopic observations must be linked with microscopic data. [Thornes (1979) provides an outstanding conceptual treatment of these process and rate relationships, relating them to the production of *form*.] Throughout the process of analysis at scales of all different magnitudes, and despite the pressure for empirical solutions, the rational analytical methods so successful historically in geomorphology (Mackin 1963) are very much needed in developing the geologically sound, man-made landscapes of the future. Much more in the way of quantitative data are needed, but the geomorphologist's perspective of *processes through time* is the most critical need for the successful design of future man-made landforms and to the selection of proper geomorphic processes in keeping with environmental goals.

## REFERENCES

Arnett, R.R. 1979. The use of differing scales to identify factors controlling denudation rates. In *Geographical approaches to fluvial processes*, A.F. Pitty (ed.), 127–47. Norwich, England: Geo Abstracts.

Baker, F.G. 1978. Variability of hydraulic conductivity within and between nine Wisconsin soil series. *Water Resources Research* 14, 103-8.

Bishop, A.W. 1966. The strength of soils as engineering materials. *Geotechnique* 16, 91–128.

Bjerrum, L. 1966. Mechanism of progressive failure in slopes of overconsolidated plastic clays and clay shales. Preprint, Am. Soc. Civil Engineers Structural Eng. Conf., Miami, Fla., 67 p.

Bradley, R.S. 1976. *Precipitation history of the Rocky Mountain States*. Boulder, Colo.: Westview Press.

Brawner, C.O. 1979. Design, construction and repair of tailings dams for metal mine waste disposal. In *Current geotechnical practice in mine waste disposal*, Committee on Embankment Dams and Slopes of the Geotechnical Engineering Division (ed.), 53–87. New York: American Society of Civil Engineers.

Brunsden, D., J.C. Doornkamp and D.K.C. Jones 1978. Applied geomorphology: a British view. In *Geomorphology: present problems and future prospects*, C. Embleton, D. Brunsden and D.K.C. Jones (eds.), 251–62. London: Oxford University Press.

Bryan, K. 1954. *Smithsonian miscellaneous collections*. Vol. 122, no. 7: *The geology of Chaco Canyon, New Mexico in relation to the life and remains of the prehistoric peoples of Pueblo Bonito*. Washington, D.C.: The Smithsonian Institution.

Chorley, R.J. 1978. Bases for theory in geomorphology. In *Geomorphology: present problems and future prospects*, C. Embleton, D. Brunsden and D.K.C. Jones (eds.), 1–13. London: Oxford University Press.

Connor, J.J., B.M. Anderson, J.R. Keith and J.G. Boerngen 1976. Soil and grass chemistry near the Four Corners powerplant. In *Geochemical survey of the western energy regions*. Open-file rept. 76–729, 112–20. Denver: U.S. Geological Survey.

de Vries, J. and T.L. Chow 1978. Hydrologic behavior of a forested mountain soil in coastal British Columbia. *Water Resources Research* 14, 935–42.

Ebens, R.J. and J.M. McNeal 1977. Geochemistry of Fort Union shale and sandstone in outcrop in the Northern Great Plains coal province. In *Geochemical survey of the western energy regions*. Open-file rept. 77–872, 185–97. Denver: U.S. Geological Survey.

Einstein, H.A. 1950. *The bed load function for sediment transportation in open channel flows*. U.S. Dept. Agriculture Tech. Bull. 1026.

Erdman, J.A. 1978. Potential toxicologic problems associated with strip mining. In *Geochemical survey of the western energy regions*. Open-file rept. 78–1105, 117–25. Denver: U.S. Geological Survey.

Everett, A.G. 1979. Secondary permeability as a possible factor in the origin of debris avalanches associated with heavy rainfall. *Jour. Hydrology* 43, 347–54.

Everett, A.G., J.J. Anderson, A.E. Peckham and L. MacMillion 1974. Engineering stability in the future spoil and waste piles of the western United States. *Geol. Soc. America Abs. with Programs* 6, 727–8.

Everett, A.G., M.S. Marshall, J.R. Kramer and H.D. Grundy 1980a. Weathering conditions in feldspathic rocks under intense rainfall conditions, Papua New Guinea. Abs., Geol. Soc. America Ann. Mtg., Atlanta.

Everett, A.G., M.L. Miller, W.C. Retzsch, F.J. Moeng and M.S. Marshall 1980b. *Estimates of geochemical processes and chemical concentrations in the Fly River drainage that may result from development of the Ok Tedi Project, Western Province, Papua New Guinea.* 2 vols. Prepared for the Department of Minerals and Energy, Papua New Guinea, under contract to Behre Dolbear and Company, Inc., New York City. Rockville, Md.: Everett and Associates.

Giesy, J.P. and L.A. Briese 1978. Particulate formation due to freezing humic wastes. *Water Resources Research* **14**, 542–4.

Groenewold, G.H. and B.W. Rehm 1980. Instability of contoured surface-mined landscapes in the Northern Great Plains: causes and implications. In *Adequate reclamation of mined lands.* Madison, Wis.: Soil Conservation Society of America.

Hack, J.T. 1942. *The changing physical environment of the Hopi Indians of Arizona.* Papers Peabody Mus. Am. Archaeology Ethnology, Harvard Univ. Vol. 35, no. 1. Repts. Awatovi Expedition Peabody Mus., Harvard Univ., Rept. no. 1.

Higgins, R.J. 1977. Monitoring and predicting the behavior of a river carrying artificially high sediment load. In *6th Australian hydraulics and fluid mechanics conference*, 109–12. Sydney: Institution of Engineers.

Kealy, C.D. and R. Busch 1979. Evaluation of mine tailings disposal. In *Current geotechnical practice in mine waste disposal*, Committee on Embankment Dams and Slopes of the Geotechnical Engineering Division (ed.), 181–201. New York: American Society of Civil Engineers.

Lamb, H.H. 1977. *Climate: present, past and future.* Vol. 2: *Climatic history and the future.* New York: Barnes & Noble Books.

Lattman, L.H. 1977. Weathering of caliche in southern Nevada. In *Geomorphology in arid regions*, D.O. Doehring (ed.), 221–31. Binghamton, N.Y.: State University of New York.

Lawson, M.P. 1974. *The climate of the great American desert.* Lincoln, Neb.: University of Nebraska Press.

Loughnan, F.C. 1969. *Chemical weathering of the silicate minerals.* New York: American Elsevier.

Mackin, J.H. 1963. Rational and empirical methods of investigation in geology. In *The fabric of geology*, C.C. Albritton, Jr. (ed.), 135–63. Reading, Mass.: Addison–Wesley.

Martin, S.C. 1975. *Ecology and management of southwestern semidesert grass-shrub ranges: the status of our knowledge.* Rocky Mountain Forest and Range Expt. Sta., USDA Forest Research Paper RM–156, Fort Collins, Colo.

McCain, J.F., L.L. Hoxit, R.A. Maddox, C.F. Chappell and F. Caracena 1979. Meteorology and hydrology in Big Thompson River and Cache la Poudre River basins, Part A. Shroba, J.F., P.W. Schmidt, E.J. Crosby and W.R. Hansen 1979. Geologic and geomorphic effects in the Big Thompson Canyon area, Larimer County, Part B. *Storm and flood of July 31–August 1, 1976, in the Big Thompson River and Cache la Poudre River basins, Larimer and Weld counties, Colorado.* U.S. Geol. Survey Prof. Paper 1115. Washington, D.C.: U.S. Government Printing Office.

Meynink, W.J.C. 1979. *Bougainville flood estimation manual.* Rept. no. ED79/HY01, Bougainville Copper Ltd., Panguna, Papua New Guinea.

Park, C.C. 1977. Man-induced changes in stream channel capacity. In *River channel changes*, K. Gregory (ed.), 121-44. New York: John Wiley.

Pickup, G. and R.J. Higgins 1979. Estimating sediment transport in a braided gravel channel - the Kawerong River, Bougainville, Papua New Guinea. *Jour. Hydrology* **40**, 283-97.

Schmidt-Collérus, J.J. 1974. *The disposal and environmental effects of carbonaceous solid wastes from commercial oil shale operations.* First ann. rept. no. NSF GI 34282X1. Washington, D.C.: National Science Foundation.

Seed, H.B. 1974. Landslides during earthquakes due to soil liquefaction. In *Terzaghi lectures 1963-1972*, J.W. Hilf (ed.), 193-261. New York: American Society of Civil Engineers.

Shen, H.W. (ed.) 1979. *Modeling of rivers.* New York: John Wiley.

Simons, D.B., J.D. Nelson, E.R. Reiter and R.L. Barkau 1978. *Flood of 31 July 1976 in Big Thompson Canyon, Colorado.* Washington, D.C.: National Academy of Sciences.

Strakhov, N.M. 1967. *Principles of lithogenesis.* Vol. 1. Edinburgh: Oliver & Boyd.

Thornes, J.B. 1977. Hydraulic geometry and channel change. In *River channel changes*, K.J. Gregory (ed.). New York: John Wiley.

Thornes, J.B. 1979. Processes and interrelationships, rates and changes. In *Process in geomorphology*, C. Embleton and J. Thornes (eds.), 91-100, 378-87. New York: John Wiley.

Vanoni, V.A. 1946. *Transport of suspended sediment by water. Am. Soc. Civil Engineers Trans.* **111**, 67-102.

# 2

# THE ROLE OF GEOMORPHOLOGY IN THE IDENTIFICATION AND EVALUATION OF NATURAL HAZARDS

*John J. Clague*

## ABSTRACT

Air photograph interpretation and detailed field investigations have been used to identify and evaluate a wide variety of natural hazards in Shakwak Valley, part of the main transportation corridor linking Alaska and western Canada. Potential natural hazards to communities, roads, and proposed major developments such as the Alaska Highway gas pipeline include (a) seismic shaking and surface rupture associated with active faults; (b) rockslides, rockfalls, slumps, and complex landslides; (c) active-layer detachment flows; (d) debris flows and debris torrents; (e) floods triggered by heavy rainfall and snowmelt; (f) jökulhlaups; (g) inundation of Haines Junction by a large glacier-dammed lake (Lake Alsek); (h) explosive volcanism; and (i) permafrost degradation. An evaluation of each of these hazards for the purpose of land use planning has been made first by determining locations of potential hazard areas relative to existing development in Shakwak Valley, and then by estimating the probability of occurrence and the likely areal extent of damage for destructive events of various magnitudes. Probabilities of the various hazards in Shakwak Valley are based on meager historical data and on radiocarbon dates on terrain features and sedimentary deposits produced by past "catastrophic" events.

The following conclusions can be drawn from estimates of the probability of occurrence and probable economic impact of natural hazards in Shakwak Valley. The greatest short- to intermediate-term (10 to 100 years) hazards to life and property are floods, debris torrents, and earthquakes. Inundation of Haines Junction by Lake Alsek and catastrophic jökulhlaups elsewhere in the valley, although potentially extremely destructive, are unlikely to occur under present climatic and glacier balance conditions. Finally, large landslides and tephra falls in Shakwak Valley not only have lower probabilities than other hazards, but also would probably cause less damage.

## INTRODUCTION

In response to a proposal by a consortium of companies to build a natural gas pipeline from Prudhoe Bay in Alaska through Canada to the midwestern United States, the Geological Survey of Canada in 1978 initiated a study of natural hazards in the vicinity of the proposed pipeline route in southwestern Yukon Territory. The area selected for detailed study was Shakwak Valley, a major trench bordering the Saint Elias Mountains (Fig. 1). This valley is part of the main transportation corridor connecting Alaska and

17

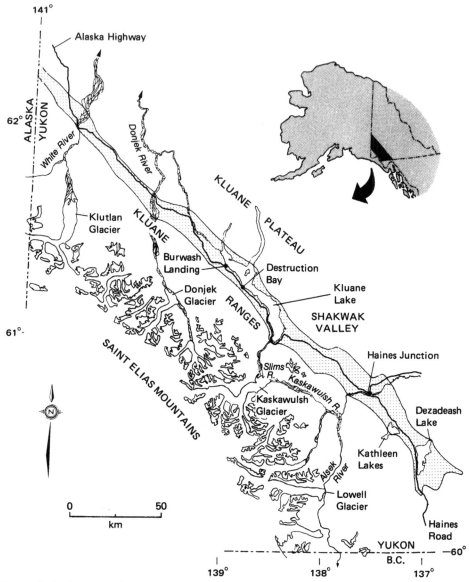

**Figure 1**  Index map of the study area.

western Canada and likely will be a locus of future economic development in the region. This area was chosen for study because of its wide variety of potential natural hazards, including most of those likely to affect development elsewhere in Yukon Territory.

## Method

The method of investigation was primarily geomorphic and involved the identification and, where possible, the dating of terrain features formed by potentially destructive natural processes. An evaluation of most major hazards for the purpose of land use planning was made by means of air photograph interpretation and subsequent field work.

First, possible hazards were identified relative to existing towns, roads, communication facilities, and other structures. Then estimates of the probability of occurrence and probable areal extent of damage for various destructive events were made on the basis of the history of past events of comparable size. Probabilities of the various hazards could not be determined completely from historical records, because reliable historical data are available for Shakwak Valley for only the last 30 to 40 years. Thus recurrence estimates for some hazards, notably those with low probabilities of occurrence, were based to a large extent on radiocarbon dates on terrain features and sedimentary deposits produced by past "catastrophic" events. Unfortunately, such dating is imprecise, and landforms and deposits often provide only a fragmentary record of past events. Nevertheless, in remote regions such as Shakwak Valley where land use decisions must be made, such information is useful and in fact may be all that is available for the evaluation of certain hazards. Despite its limitations, this information often is sufficient for first-order estimates of the probability of future destructive events.

In this paper the main natural hazards in Shakwak Valley are first catalogued and then evaluated in terms of their probable future impact on life and property. An exhaustive summary of all hazards in this region is not possible because of the reconnaissance nature of the study and limitations on the length of the paper. Instead, the main hazards are briefly summarized, largely to show the role of geomorphology in the identification and evaluation of hazards. Hazards due exclusively to man's activities and those which are not controlled by geological or geomorphic factors are excluded from this paper. For example, forest fires, although potentially destructive to structures in much of the study area, are not geologically or geomorphically controlled and thus are not considered.

## Setting

Shakwak Valley is a structural trough about 4 to 20 km wide trending northwest-southeast across southwestern Yukon Territory (Fig. 1). Its floor is characterized by gently rolling and glacially fluted topography, locally incised by streams. Shakwak Valley is bounded on the northeast by the Kluane Plateau, which consists mainly of rounded peaks and broad undulating ridges separated by broad valleys; it is bounded on the southwest by the Kluane Ranges, which comprise steep slopes and serrated narrow ridges and peaks (Bostock 1948). The Kluane Ranges form the easternmost block of the Saint Elias Mountains, an extremely rugged high-relief massif with large ice fields and valley glaciers feeding numerous high-gradient streams and rivers. The northeastern flank of the Kluane Ranges, although breached in places by rivers, rises from Shakwak Valley as a spectacular wall, with relief locally exceeding 1600 m. In contrast, the opposing northeast side of the valley rises irregularly, in places interrupted by steep bluffs and in others by subdued hills, to the level of the Kluane Plateau surface.

Shakwak Valley is underlain by thick unconsolidated sediments, chiefly till, outwash, and glaciolacustrine silt and clay deposited during various Pleistocene glaciations, and gravelly alluvium and peat deposited during Holocene time (Hughes et al. 1972). Sediments on mountain slopes and upland areas bordering Shakwak Valley, on the other hand, are relatively thin and consist mainly of till and colluvium (Rampton 1979a–1981). There are also large areas of rock outcrop on steep mountain slopes. In the eastern Kluane Ranges, bedrock consists mainly of intensely faulted Paleozoic and Mesozoic sedimentary and volcanic rocks (Campbell & Dodds 1978). Bedrock underlying the Kluane Plateau east of Shakwak Valley consists mainly of early Paleozoic and late Precambrian high-grade metamorphic rocks intruded by Mesozoic and early Tertiary plutons (Muller 1967; Tempelman–Kluit 1974).

Shakwak Valley is an important transportation and communication corridor, although it presently is sparsely habitated. The Alaska Highway and Haines Road, major road links connecting Alaska and Canada, are located in the valley (Fig. 1). The resident population of the valley is about 600 to 700, largely concentrated in the towns of Haines Junction, Destruction Bay, and Burwash Landing [Foothills Pipe Lines (South Yukon) Ltd. 1979]. Haines Junction, with a population of about 300 to 500, is slated for major growth as the headquarters of Kluane National Park and as a maintenance and supply center for the proposed Alaska Highway gas pipeline.

## NATURAL HAZARDS

Potential natural hazards to communities and roads in Shakwak Valley and to proposed major developments such as the gas pipeline include (a) seismic shaking and surface rupture associated with active faults; (b) rockslides, rockfalls, slumps, and complex landslides, many probably earthquake-induced; (c) active-layer detachment flows on slopes underlain by permafrost; (d) debris flows and debris torrents; (e) floods caused by excessive rainfall and snowmelt and accompanied by bank erosion and channel shifting; (f) jökulhlaups; (g) formation of a large glacier-dammed lake (Lake Alsek) capable of inundating Haines Junction; (h) explosive volcanism resulting in air-fall deposition of tephra in Shakwak Valley; and (i) permafrost degradation related to construction activities and pipeline operation.

### Seismicity and faulting

The Kluane Ranges and Shakwak Valley are located near the eastern edge of an area of high seismicity related to the interaction of the Pacific and North American crustal plates. The Fairweather Fault, which is only about 100 km to the southwest of Shakwak Valley, is thought to be the present transform boundary between the two plates and is characterized by large earthquakes accompanied by surface rupturing along much of its length (Plafker et al. 1978). Other major faults occur in the Saint Elias Mountains and Shakwak Valley. One of the largest of these is the Denali Fault, which follows the western edge of Shakwak Valley from the Alaska–Yukon boundary to the south end of Kluane Lake (Campbell & Dodds 1978). The Denali Fault is a major intracontinental crustal break extending for more than 2000 km across south–central and southeastern Alaska, southwestern Yukon Territory, and northern British Columbia, with Cenozoic displacements perhaps as large as 350 km (Eisbacher 1976; Lanphere 1978). Because some segments of the fault in Alaska are presently active (e.g., Richter & Matson 1971; Hickman & Craddock 1973; Stout et al. 1973; Hickman et al. 1977; Plafker et al. 1977), the possibility that the segment of the fault in Shakwak Valley is also active must be considered in any evaluation of natural hazards: Clague (1979a) surveyed raised beaches of Kluane Lake which are younger than 500 years old and found that they are not offset where crossed by the Denali Fault. Although the fault cuts glacial landforms of latest Pleistocene age in Shakwak Valley, and probably cuts early or middle Holocene river terraces and fans, it does not offset late Holocene floodplains and alluvial fans (Fig. 2). It thus appears that there has been little or no surface rupturing on the Denali Fault in

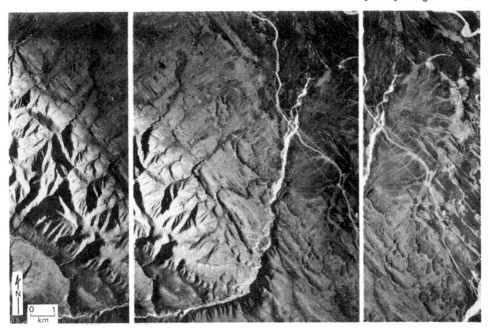

**Figure 2** Photo stereogram showing a late Quaternary fault scarp in Shakwak Valley northwest of Burwash Landing. The scarp, which is indicated by arrows, cuts late Pleistocene glacial landforms, but not modern floodplains and alluvial fans. Early to middle Holocene terraces apparently are offset at the site marked "x." Photo numbers A15739–50, -51, and -52 (National Air Photo Library, Canada Department of Energy, Mines and Resources).

Yukon Territory during the last few hundred to several thousand years. However, it must be emphasized that minor prehistoric displacements, on the order of 0·5 m, probably would not be detected in this region with the geomorphic investigative approach used in this study.

Although the Denali Fault in Yukon Territory has been active to only a minor extent during late Quaternary time, present-day seismicity in the region is high (Fig. 3). Several moderate earthquakes (M 5–6+) have occurred in the Shakwak Valley/Kluane Ranges area since 1899 when earthquake data collection began. Patterns of both historic seismicity and microearthquake distribution (Boucher & Fitch 1969; R.B. Horner, personal communication, 1980), indicate that most earthquakes in the region occur in an intensely deformed and faulted belt between the Denali and Duke River Faults in the Saint Elias Mountains. The most striking geomorphic evidence of this seismicity is sharp, uphill-facing scarps and associated trenches which are widespread in the Kluane Ranges (Fig. 4). Most of these scarps are the surface traces of faults produced by deep-seated gravitational spreading and subsidence during major earthquakes (Eisbacher & Hopkins 1977; Clague 1979a). Although these features cannot be dated, some are fresh and little modified by erosion and probably have formed in the last few hundred years. They indicate that large, potentially destructive earthquakes can be expected in the future in the mountains west of Shakwak Valley. It has been estimated that ground accelerations of 0·1 g – the approximate threshold of accelerations capable of damaging ordinary struc-

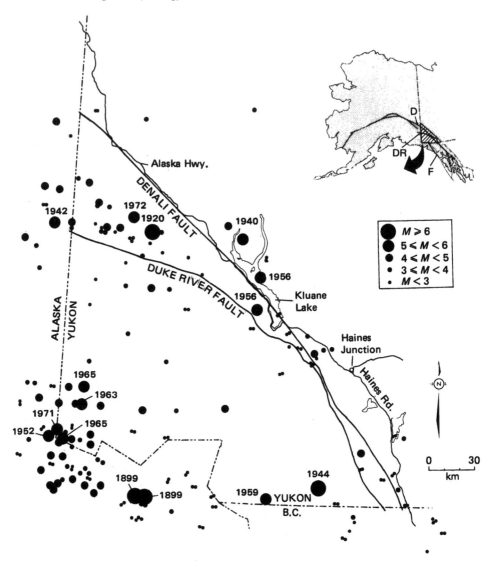

**Figure 3** Earthquake epicenters in southwestern Yukon Territory and bordering parts of British Columbia and Alaska, 1899–1978. Dates of larger earthquakes are indicated. Also shown are major fault systems, including the Denali (D), Duke River (DR), and Fairweather (F) Faults (see index map, Fig. 1). Earthquakes that occurred prior to the establishment of the Whitehorse seismograph station in 1971 have epicentral locations accurate only to several tens of kilometers. Epicenters of earthquakes occurring after 1971 are accurate to within about 20 km. Consistent detection of seismic events greater than magnitude 4 has been possible only since 1971, and events greater than magnitude 5 since 1964. All plotted earthquakes smaller than magnitude 3 occurred in 1978 and are more accurately located due to recent improvements in the seismograph network. Data supplied by Division of Seismology and Geothermal Studies, Canada Earth Physics Branch.

tures on firm soil – have a return period of 30 to 100 years for the Shakwak Valley area (Stevens & Milne 1974). Although this estimate undoubtedly will be revised somewhat in the future due to recent improvements in the seismograph network in southwestern Yukon Territory, it does indicate the potential for significant earthquake damage in Shakwak Valley.

## Landslides

Landslides are abundant on moderate and steep slopes in the Saint Elias Mountains. Large rockslides, slumps, and complex landslides are most common in areas of Tertiary rocks (Fig. 5); abundant small rockfalls and rockslides occur in all bedrock terrains, their locations being largely controlled by topography and bedrock stratification and structure. Many of these failures probably are earthquake triggered. Other types of landslides that are widespread in the Saint Elias Mountains are debris flows and active-layer detachment flows. The former commonly occur during intense precipitation events on steep colluvial or till-covered slopes. A relatively thin layer of weathered saturated colluvium fails and becomes mobilized into a flow which advances down a gully to the foot of the slope where mobility is lost. Active-layer detachment flows are shallow failures on moderate

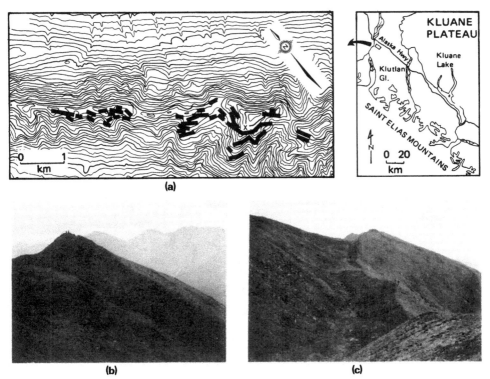

**Figure 4** Gravitational ridge spreading in the Kluane Ranges. (a) Locations of uphill-facing scarps and associated trenches on a ridge bordering Shakwak Valley. Topographic contour interval is 30 m (100 ft). (b and c) Photographs of an uphill-facing scarp (site "x" in Fig. 4a).

slopes blanketed by medium- to fine-textured sediments and underlain by permafrost. They are similar to debris flows except that failure occurs at the base of the active layer. Landslides in the Saint Elias Mountains range widely in age. Some apparently formed during deglaciation at the close of the Pleistocene, because they bear morphologic features indicative of deposition on or against ice. Others are very fresh and appear to have formed in the last several centuries.

Large landslides are uncommon in Shakwak Valley and occur only on the southwest side of the valley beneath steep slopes of the Kluane Ranges (Fig. 5). Several landslides, traversed by the Alaska Highway near the south end of Kluane Lake, were studied in

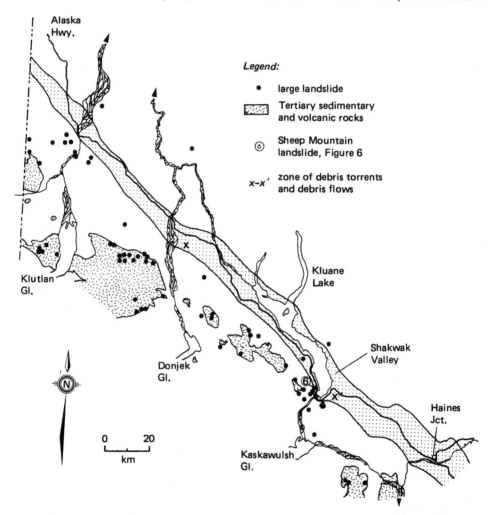

**Figure 5**  Large landslides in parts of the Kluane Ranges, Shakwak Valley, and Kluane Plateau. Many of these landslides are in Tertiary sedimentary and volcanic rocks. Debris flows and debris torrents in Shakwak Valley are most common on an alluvial apron fronting the Kluane Ranges between Donjek River and the south end of Kluane Lake (zone x–x'). Distribution of Tertiary rocks from Campbell and Dodds (1978), and landslides from Rampton (1979a–1981).

detail in an effort to determine their mechanisms and ages, information that could be used to estimate the likelihood of future slope failures in the area. The landslide shown in Figure 6, for example, formed when a steep face of highly fractured metavolcanic rocks on the flank of the Kluane Ranges collapsed, causing debris to stream down a gully and

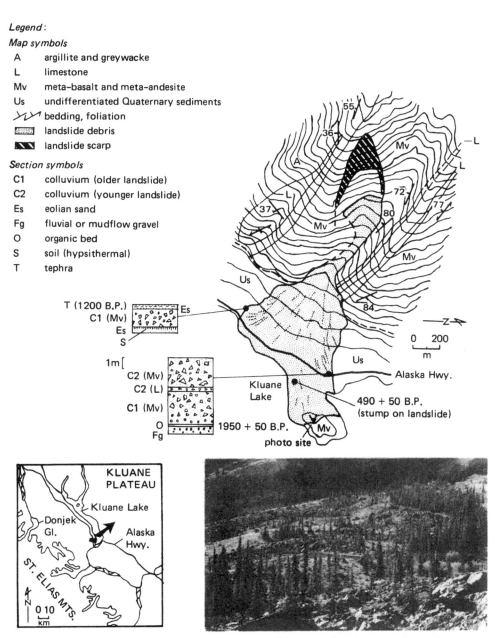

*Legend*:

*Map symbols*

| | |
|---|---|
| A | argillite and greywacke |
| L | limestone |
| Mv | meta–basalt and meta–andesite |
| Us | undifferentiated Quaternary sediments |
| ⟩↙↗ | bedding, foliation |
| ▦ | landslide debris |
| ◤◥ | landslide scarp |

*Section symbols*

| | |
|---|---|
| C1 | colluvium (older landslide) |
| C2 | colluvium (younger landslide) |
| Es | eolian sand |
| Fg | fluvial or mudflow gravel |
| O | organic bed |
| S | soil (hypsithermal) |
| T | tephra |

**Figure 6** The Sheep Mountain landslide, a two-phase rockslide crossed by the Alaska Highway near the south end of Kluane Lake. Topographic contour interval is 30 m (100 ft). Bedrock geology modified from Read and Monger (1976).

across an alluvial fan in Shakwak Valley. Blocks in excess of 1500 tonnes were transported in the rapidly moving slide mass. The stratigraphy of the slide debris as exposed
in road cuts along the Alaska Highway indicates that the landslide is composite and
consists of at least two superposed slide units. An older phase of landsliding occurred
after 1950 ± 50 years B.P. (GSC-2850), which is the age of wood in a forest bed underlying slide debris; it occurred earlier than about 1200 years B.P., the age of White River
tephra, which overlies the lower slide unit. The younger phase apparently postdates the
White River tephra, which is absent from the landslide surface, but is older than 490 ± 50
years B.P. (GSC-2860), the age of the pith of a dead stump rooted on the landslide.
Other landslides in this area also consist of units both younger and older than the White
River tephra. However, the surfaces of these landslides are partially covered by Neoglacial
loess derived from the nearby Slims River floodplain and deposited in part in the last
several hundred years (Denton & Stuiver 1966; Hughes et al. 1972). Thus despite the
relative abundance of landslides at the south end of Kluane Lake, there has been little
significant mass movement in this area during the last few centuries. This indicates that
the probability of future large landslides here and elsewhere in Shakwak Valley during the
next century is low.

Relatively small debris flows and active-layer detachment flows are comparatively
common in the Kluane Ranges and on some sediment-covered slopes bordering Shakwak
Valley (e.g., Broscoe & Thomson 1969). A typical example is a debris flow that blocked
the Alaska Highway at the south end of Kluane Lake following a period of intense rainfall
in the summer of 1967 (Hughes et al. 1972). The flow originated near the headwall of a
small ravine 800 m higher than the highway, then proceeded down the ravine and across
a small fan at the foot of the mountain slope. The flow moved boulders up to 2·4 m in
diameter and locally built levees 1 to 1·5 m high. Large active alluvial fans fronting the
Kluane Ranges in Shakwak Valley are sites of debris-flow and debris-torrent activity (Fig.
5). Debris flows mainly affect the apices of these fans, whereas more fluidized debris
torrents are carried much farther down the fans from valley mouths. Small active-layer
detachment flows are most common on moderate slopes underlain by till and White
River tephra north of Kluane Lake. South of Kluane Lake, permafrost is less extensive
on low slopes bordering Shakwak Valley; consequently such flows are rare there.

Large debris flows, debris torrents, or floods might result if a stream flowing into
Shakwak Valley became blocked by a landslide in the Kluane Ranges. The lake ponded
behind such a landslide might overtop and rapidly incise the landslide debris, or it might
cause the debris dam to collapse. In either case, it is likely that a large slurry of sediment
and water would be carried into Shakwak Valley. Although no such catastrophic events
have occurred in historical times in the region, and none can be documented on geomorphic grounds, the possibility of a future occurrence cannot be ruled out.

Finally, there is the potential for landsliding at the delta front of Slims River at the
south end of Kluane Lake. The floodplain and delta of Slims River are underlain by thick
clayey silt of medium to high plasticity (R.M. Hardy & Associates Ltd. 1978; Terrain
Analysis and Mapping Services Ltd. 1978). Undisturbed surface sediments on the delta
front and on the adjacent floor of Kluane Lake have high water contents, generally 70 to
100% by weight, and high liquidity indices. These sediments thus are sensitive and subject
to loss of strength or competency when disturbed. A major earthquake centered in the
vicinity of the Slims River delta conceivably could trigger large subaqueous liquefaction
slides which might sever the Alaska Highway, although no estimate of probability can be
made for such an event.

# Floods

Floodplains and low terraces adjacent to all streams are subject to flooding (Fig. 7). Although there are no long-term water discharge records for any of these streams, field observations indicate that there are very large discharge variations caused by summer storms and by large seasonal differences in the supply of meltwater from snowpack and glaciers. Extreme discharge variability is reflected geomorphically by multiple distributary channels on alluvial fans and aprons and by poorly vegetated braided floodplains of the larger streams and rivers (Fig. 8). It is estimated that flooding causing at least localized damage occurs on the average once every 5 to 20 years.

A related hazard is bank erosion and channel shifting. At high discharges a stream entering Shakwak Valley may erode its banks and channel, or may abruptly shift its channel from one part of the floodplain to another. Channels on the wide braided floodplains of White, Donjek, Kaskawulsh, and Slims Rivers are constantly shifting, owing in part to high sediment loads, large fluctuations in discharge, and winter aufeis develop-

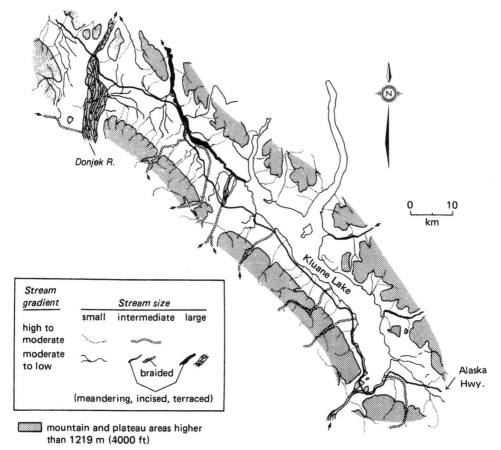

**Figure 7** Major streams subject to flooding in part of Shakwak Valley. Among the most hazardous streams are the large braided rivers flowing from glaciers in the Saint Elias Mountains and the relatively small high-gradient creeks crossing fans and aprons at the front of the Kluane Ranges.

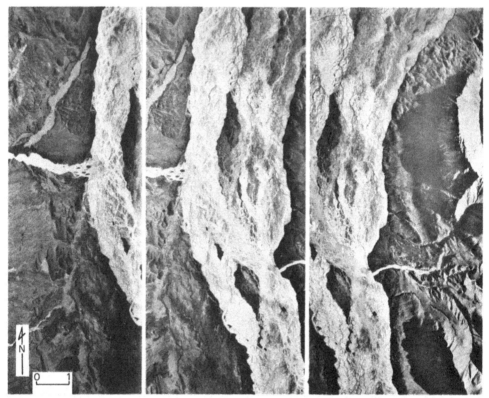

**Figure 8**   Photo stereogram of Donjek River, one of the large braided rivers flowing into Shakwak Valley from the Saint Elias Mountains. This river is characterized by very large seasonal variations in discharge and by frequent shifts in its channel pattern. Photo numbers A15728-54, -55, and -56 (National Air Photo Library, Canada Department of Energy, Mines and Resources).

ment. Streams flowing from the Kluane Ranges across alluvial fans and aprons also are unstable during periods of high discharge. Attempts have been made to constrain the flow of the most unruly of these streams to single channels on the fan surfaces, but the possibility of serious localized flooding still remains.

Although White, Donjek, Kaskawulsh, and Slims Rivers exhibit large diurnal and seasonal differences in discharge due to variations in snow- and ice-melt, the rare large floods along these rivers resulting from the rapid draining of glacier-dammed lakes in the Saint Elias Mountains are perhaps more significant as natural hazards (Glaciology Division 1977; Clague 1979b). Within the drainage basins of these rivers are both small existing glacier-dammed lakes and large potential lakes that might form in the event of a glacier advance (Fig. 9). Hazard Lake (ca. $14 \times 10^6$ m$^3$ in volume) in the Donjek River basin is one of the largest of the existing lakes. It drained abruptly in July 1975, in July or August 1977, and again in the late summer of 1978 (Collins & Clarke 1977; R.W. May, personal communication, 1979). In each case there was apparently no flooding of the Alaska Highway at the Donjek River crossing 67 km from the lake. This is due perhaps to the fact that the jökulhlaups became attenuated as they progressed down Steele Glacier

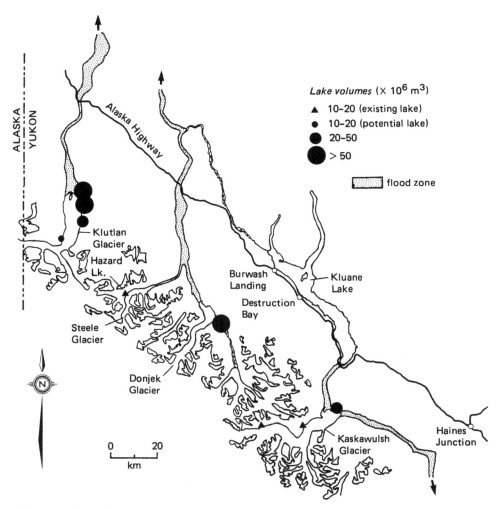

**Figure 9** Jökulhlaup zones and existing and potential glacier-dammed lakes larger than $10 \times 10^6$ m³ within the White, Donjek, and Slims-Kaskawulsh basins in Yukon Territory. Potential lakes are basins in which water would be ponded in the event of a major glacier advance. Basins near the termini of Klutlan and Kaskawulsh Glaciers have not held water for several centuries and are unlikely to do so in the foreseeable future.

and the broad Donjek River floodplain. Furthermore, theoretically Hazard Lake is capable of producing a peak flood discharge of no more than 440 m³/s (Table 1), which is within the range of normal summer discharge maxima of Donjek River. Thus it is unlikely that jökulhlaups from other existing glacier-dammed lakes, which are all smaller than Hazard Lake or of comparable size, could cause damage in Shakwak Valley unless they occurred during freshet or during a period when river channels were constricted or plugged by ice.

Much larger, potentially destructive floods might result if Klutlan, Donjek, or Kaskawulsh Glacier advanced to impound new glacier-dammed lakes (Fig. 9). For example,

Table 1 Comparison of normal discharge range of White, Donjek, and Slims Rivers and possible maximum jökulhlaup discharges in the valleys of these rivers

| River | Normal discharge (m³/s)[a] | | | Largest existing lake in basin[b] | | Largest potential lake in basin[b] | |
|---|---|---|---|---|---|---|---|
| | Max. | Min. | Mean | Max. jökulhlaup (m³/s)[c] | Distance to Alaska Hwy. (km) | Max. jökulhlaup (m³/s)[c] | Distance to Alaska Hwy. (km) |
| White | 1126 | 14 | 112 | 270 | 76 | 2186 | 49 |
| Donjek | – | – | – | 440 | 67 | 2268 | 64 |
| Slims | 320 | <1 | – | 440 | 60 | 776 | 14 |

[a] Data for White River from Water Survey of Canada records, 1975–78. No streamflow records are available for Donjek River, but discharge values are probably somewhat lower than those of White River (Glaciology Division 1977, p. 20–22). Data for Slims River from miscellaneous measurements reported by Fahnestock (1969). Max., maximum instantaneous discharge; min., minimum daily discharge; mean, mean discharge for period of record.

[b] Data from Glaciology Division (1977).

[c] Jökulhlaup discharges determined from empirical equation of Clague and Mathews (1973): $Q = 75V^{0.67}$, where $Q$ is the maximum instantaneous discharge and $V$ is the lake volume.

an advance of the toe of Donjek Glacier of less than 1 km during a surge or due to a more positive glacier regimen might block Donjek River and form a lake with a maximum volume of about $162 \times 10^6$ m$^3$. Theoretically, a flood with a peak discharge of about 2270 m$^3$/s might result from the failure of the ice dam impounding this lake (Table 1), although the flood undoubtedly would attenuate substantially before reaching Shakwak Valley. Shorelines and lacustrine sediments in Donjek Valley immediately upvalley of the terminus of Donjek Glacier indicate that such a lake has formed and drained several times since the glacier reached its Neoglacial maximum 300 to 440 years ago (Denton & Stuiver 1966; M.S. Perchanok, personal communication, 1979). In contrast, in order for Klutlan Glacier to impound large ice-marginal lakes, it must advance to positions last attained 300 to 1100 years ago (Rampton 1970). Although the Klutlan is a surging glacier, minor surges probably would not affect the position of the terminus, because movement would be damped by the massive ice-cored Neoglacial moraines extending down the valley from the active front of the glacier. A large surge might cause a significant advance of the Klutlan terminus, but the past history of Neoglacial fluctuations of the glacier indicates that an advance large enough to pond ice-marginal lakes at the sites shown in Figure 9 is unlikely to occur under existing climatic conditions. Similarly, the largest potential lake associated with Kaskawulsh Glacier has a small probability of forming under the present climatic regime. Thus although the potential floods from such lakes are very large indeed, the likelihood of their occurring in the next several decades is low, except possibly in Donjek River basin.

## Neoglacial Lake Alsek

Several times during the late Holocene, Lowell Glacier, a large surging glacier in the Saint Elias Mountains, advanced across Alsek Valley and blocked south-flowing Alsek River (McConnell 1905; Kindle 1952; Johnson & Raup 1964; Hughes et al. 1972; Clague 1979b; Rampton 1981). The resulting lake backed up into Shakwak Valley in the vicinity of Haines Junction (Fig. 10). Parts of what are now the Alaska Highway and Haines Road, the site of Haines Junction, and part of the proposed Alaska Highway gas pipeline route were inundated during high stands of the lake. At its maximum, Lake Alsek was at least 110 km long and 200 m deep at the glacier dam, and thus was the largest known Holocene glacier-dammed lake in North America.

The past history of Lake Alsek was determined to assess the likelihood of future ponding events which might cause damage in the Haines Junction area. Studies of Alsek beaches, wave-cut benches, lake deposits, and accumulations of driftwood (Figs. 11 and 12) indicate that the lake has had many ponding phases, each separated by an interval during which the present southward drainage pattern in Alsek Valley prevailed (Clague 1979b; Rampton 1981). The evidence for this includes (a) multiple weak soils and organic horizons in Alsek lacustrine deposits, and (b) multiple driftwood layers and co-incident breaks in the density and maximum size of lichens on gravelly and rubbly beach deposits at various levels in the basin. Each of the lake phases perhaps consisted of many short-lived cyclic filling and draining events, as is common for present-day, self-dumping, glacier-dammed lakes (Stone 1963; Mathews 1965, 1973; Post & Mayo 1971). Drainage occurred as a result of the failure of the ice dam at Lowell Glacier. The presence of flood bed forms, including giant bars and dunes, on the floor of Alsek Valley both upstream and downstream of Lowell Glacier attests to the rapidity with which the lake drained and to the forces exerted on the valley floor by the moving water (Fig. 12).

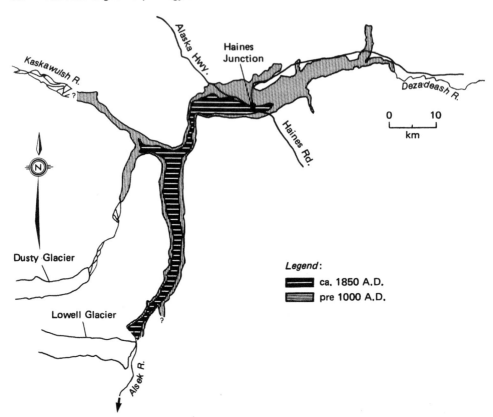

**Figure 10** Extent of Lake Alsek about A.D. 1850 (filling limit = 595 m elevation) and earlier during Neoglacial time (671 m). Although its maximum extent is not well established, Neoglacial Lake Alsek probably never exceeded about 678 m in elevation. The extent of early phases of the lake in upper Kaskawulsh Valley is unknown because there has been extensive aggradation in the valley near the toe of Kaskawulsh Glacier during the last several centuries.

Radiocarbon dating of driftwood and buried organic horizons indicates that there have been several phases of Lake Alsek in the last 500 years, as well as one or more older Neoglacial episodes of ponding. The last and least extensive expansion of the lake into the Haines Junction area occurred during the 19th century. From the age of the oldest living trees on the floor of Alsek Valley, Kindle (1952) estimated that Lake Alsek last occupied Shakwak Valley about A.D. 1850. Rampton (1981) made a similar estimate using lichen growth-rate estimates and tree-ring counts. Finally, dendrochronological analyses of driftwood indicate that this lake phase occurred sometime after A.D. 1848. Support for these age estimates is provided by Indian legends of a catastrophic flood on the Alsek River delta downstream from Lowell Glacier in southeastern Alaska in A.D. 1852 or shortly thereafter (de Laguna 1972, p. 276). Indians attributed this flood to the "breaking of a glacier" that crossed Alsek Valley.

The past history of Lake Alsek suggests that future ponding events are likely. The terminus of Lowell Glacier presently is less than 2 km from the east wall of Alsek Valley;

**Figure 11** Beaches (a), wave-cut benches (b), and driftwood accumulations (c, d, and e) of Lake Alsek.

thus during a major surge the glacier probably would block the river and create a new lake. It is not known, however, how large the resulting lake might grow. This depends in part on the present and future regimen of Lowell Glacier and on the magnitude of the surge blocking the valley. Because Lowell Glacier has receded and thinned during the last few centuries (Rampton 1981), it is unlikely that it can now impound a lake comparable in size to the large lakes existing in Alsek Valley prior to the ca. A.D. 1850 ponding phase.

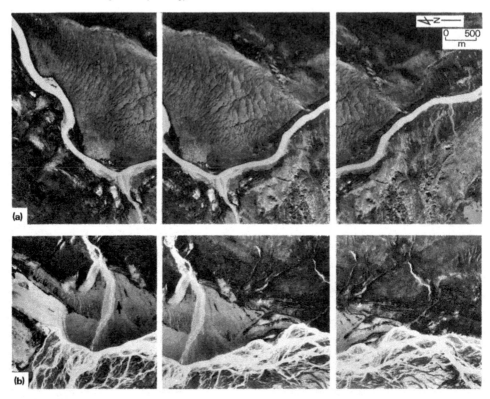

**Figure 12** Stereograms illustrating Lake Alsek dunes and beaches. (a) Large dunes on the floor of Alsek Valley immediately upstream of the terminus of Lowell Glacier (some of the Neoglacial moraines of Lowell Glacier are visible in the right stereopair). Photo numbers A23819–48, –51, and –52. (b) Beaches and dune remnants in Alsek Valley near the confluence of Dezadeash and Kaskawulsh Rivers. A driftwood layer deposited on the shore of Lake Alsek about A.D. 1850 is indicated by an arrow. Photo numbers A23793-198, –199, and –200. All photos from National Air Photo Library, Canada Department of Energy, Mines and Resources. Scale and north direction shown on Figure 12a are the same in Figure 12b.

## Volcanism

Twice in the last two millenia there have been large explosive volcanic eruptions from a vent in the Saint Elias Mountains near the Alaska–Yukon boundary. Tephra generated during these eruptions is termed the White River ash and covers a large area of southern Yukon Territory and eastern Alaska (Fig. 13; Capps 1916; Bostock 1952; Berger 1960; Stuiver et al. 1964; Lerbekmo & Campbell 1969; Rampton 1969; Hughes et al. 1972; Lerbekmo et al. 1975). This tephra is about 100 m thick at the vent and exceeds 50 cm in thickness in parts of Shakwak Valley (Bostock 1952). Radiocarbon dates on organic material associated with two discrete layers of White River ash indicate that eruptions occurred about 1200 years ago and 1500 to 1800 years ago (Hughes et al. 1972). Tephra produced by the earlier eruption was deposited in a lobe extending north from the vent;

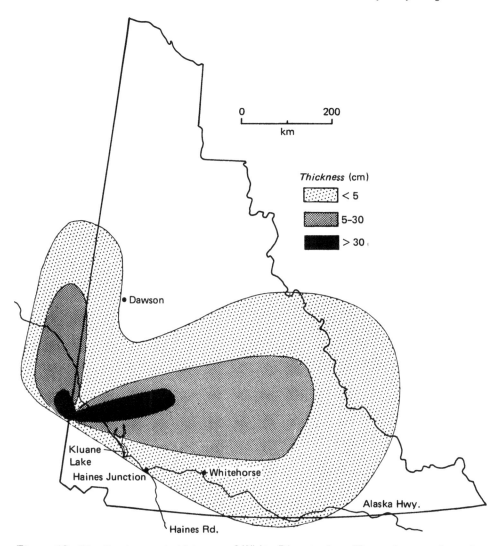

**Figure 13** Distribution and thickness of White River tephra. The tephra consists of an older northerly trending lobe and a younger easterly trending lobe. Isopach map modified from Bostock (1952), Lerbekmo and Campbell (1969), Rampton (1969), and Hughes et al. (1972).

In contrast, the younger tephra occurs in a lobe extending east from the vent (Fig. 13). A future eruption in the Saint Elias Mountains comparable to those which produced the White River ash likely would cause damage in Shakwak Valley. Aside from direct damage that might result from tephra fallout, the drainage pattern of the White or other rivers might be disrupted by volcanic debris, with resulting flooding in Shakwak Valley.

The future eruptive history of the Saint Elias Mountains cannot be predicted in detail. However, the past history of volcanism summarized above suggests that the probability of an eruption causing significant damage in Shakwak Valley in the next few centuries is low.

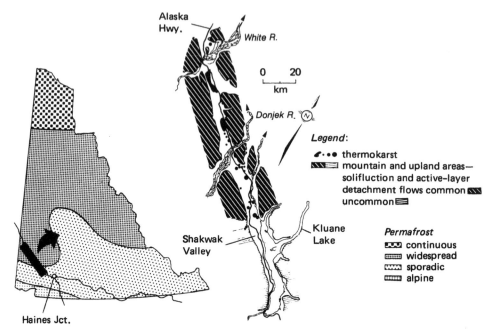

**Figure 14** Generalized distribution of permafrost in Yukon Territory (after Brown 1967) and permafrost hazards in northern Shakwak Valley.


## Permafrost

Shakwak Valley lies within the zone of discontinuous permafrost (Fig. 14; Brown 1967). In general, permafrost in this region decreases in extent from north to south, from high elevations to low, and is more common on north-facing slopes than south-facing slopes. Thus whereas permafrost underlies almost all of northern Shakwak Valley, with the exception of large lakes and river channels, it is absent over much of the southern part of the valley and on adjacent south-facing slopes at low elevations. Permafrost and ground ice are best developed in poorly drained fine-grained sediments and peat. Ground ice occurs in the form of ice lenses and veinlets and ice wedges. Differential melting of this ice has produced many shallow, flat-bottomed lakes in northern Shakwak Valley (Fig. 14), some of which are presently expanding. Active thermokarst terrain represents an obvious hazard to future development; however, this hazard can be mitigated through appropriate engineering design [Foothills Pipe Lines (South Yukon) Ltd. 1979] or, alternatively, by avoiding areas subject to thermokarst expansion.


## DISCUSSION

The foregoing indicates that a wide variety of natural hazards exist in Shakwak Valley. However, the future danger to life and property posed by each of these hazards differs for at least three reasons: (a) prior mitigative action is feasible for some hazards, but not for others; (b) the probability of occurrence of each hazard is different, as is (c) the likely areal extent and severity of damage.

Mitigative measures may be taken to minimize the danger of damage to roads, buildings, pipelines, and other structures resulting from permafrost degradation. An extensive shallow drilling program in Shakwak Valley has been conducted by Foothills Pipe Lines (South Yukon) Ltd. to determine, among other things, the extent and character of permafrost along the proposed Alaska Highway gas pipeline route. Foothills plans to adopt design guidelines to ensure minimal permafrost disturbance during pipeline construction and operation [Foothills Pipe Lines (South Yukon) Ltd. 1979]. Areas especially sensitive to permafrost thaw, including thermokarst terrain, will probably be avoided. Mitigative or remedial action also may be taken in the event that Lake Alsek forms again and threatens Haines Junction. It would take about one year for the lake to grow large enough to inundate parts of Haines Junction and the Haines Road (G.K.C. Clarke, personal communication, 1979). Presumably, this would be enough time to relocate low-lying portions of the community and reroute the road, or, alternatively, to attempt to remove the ice dam at the toe of Lowell Glacier or construct a drainage bypass for Alsek River. However, the feasibility of artificially draining a lake dammed by flowing ice 5 km wide and several hundred meters thick has yet to be shown. Furthermore, the environmental and political consequences of such an undertaking in an otherwise undisturbed, sensitive part of a national park would be considerable. In any case, mitigative or remedial action will be disruptive and costly if Lake Alsek forms again and grows to beyond its mid-19th century size. Mitigative measures to minimize earthquake, flood, and landslide damage probably are economically feasible only for large expensive developments such as the proposed gas pipeline. Existing development in Shakwak Valley can be expected to suffer some damage from earthquakes, floods, and landslides, the amount depending largely on the location and magnitude of the destructive event.

The probabilities of threshold destructive events for most natural hazards in Shakwak Valley are compared in Figure 15. These probabilities are based on estimates of the past recurrence of events of comparable size, and take into account the disposition of existing, not future, buildings, roads, bridges, and communications facilities. The figure indicates that floods (other than jökulhlaups), debris torrents, debris flows, and earthquakes are the most probable destructive events in Shakwak Valley; jökulhlaups, inundation by Lake Alsek, and large landslides are less likely; and volcanism has a very low probability. The natural hazards differ not only in terms of probability of occurrence, but also in the size

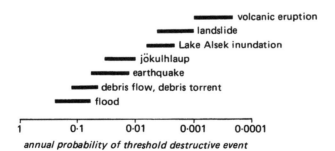

**Figure 15** Probability of threshold destructive events for various natural hazards in Shakwak Valley. Permafrost is not included because there are insufficient data to evaluate the probability of future damage to existing buildings and roads due to melting of ground ice.

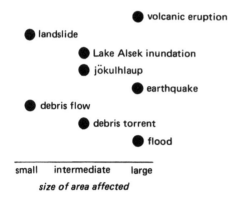

volcanic eruption
landslide
Lake Alsek inundation
jökulhlaup
earthquake
debris flow
debris torrent
flood

small    intermediate    large
*size of area affected*

**Figure 16** Areal extent of damage resulting from large destructive events in Shakwak Valley. Damage caused by floods and debris torrents would be restricted to floodplains and fans, but many of these sites within a relatively large area probably would be affected.

of the area in Shakwak Valley likely to be affected (Fig. 16). For example, a large snowmelt flood or a major earthquake would probably affect a relatively large region, whereas a landslide would have only localized effects.

An evaluation of the relative importance of each of the identified natural hazards may be made of combining estimates of probability (Fig. 15) with estimates of areal extent of damage (Fig. 16). The result of such an analysis is shown in Figure 17, where anticipated damage to existing buildings, roads, and bridges in Shakwak Valley is compared for various hazard events with return periods of 10, 100, and 1000 years. Because of the limitations of the data base, damage is here expressed in a relative, rather than absolute sense, and specific dollar values are not affixed to the damage estimates. It also should be emphasized that the estimates apply only to Shakwak Valley and its bounding walls, not to the Kluane Plateau or Kluane Ranges proper. This is worth pointing out because the probabilities of some hazards vary according to physiographic region. For example, rockslides and rockfalls are relatively common in the Kluane Ranges, but are rare in Shakwak Valley. There are also differences in the likelihood of various hazards within the valley itself. For example, rockslides and rockfalls are most likely to occur along the steep-walled, southwest side of the valley (Fig. 5); jökulhlaups can occur only on the White, Donjek, and Slims River floodplains (Fig. 9); and Lake Alsek inundation will be limited to the Haines Junction region (Fig. 10).

With these comments in mind and turning to Figure 17, we see that damage attributable to the major identified natural hazards is likely to be low in the short term (10-year time frame). This, at first glance, might seem unusual in light of the rather imposing set of potential hazards catalogued for the region. However, present development in Shakwak Valley is relatively limited, and many of the most hazardous areas (e.g., steep valley walls, floodplains, active fans) thus far have, for the most part, been avoided. If anticipated economic development of the valley proceeds, it is probable that increased short-term damage will ensue, largely from floods and debris torrents, but possibly also from debris flows and earthquakes. In the intermediate term (100-year time frame), moderate to extensive damage to existing roads and bridges will probably result from floods, debris torrents, and earthquakes; damage to existing buildings, except in the case of a 100-year earthquake, will be considerably less. Inundation of Haines Junction by Lake Alsek and catastrophic jökulhlaups, although potentially extremely destructive, are unlikely to occur under present climatic and glacier balance conditions; thus damage attributable to them in the short and intermediate term is relatively low. Finally, large

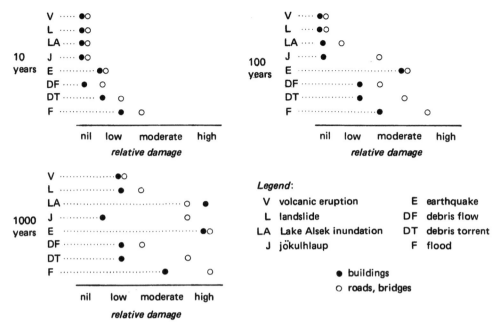

**Figure 17** Estimates of future relative damage to existing buildings, roads, and bridges in Shakwak Valley for various natural hazards. Anticipated damage is plotted for events with return periods of 10, 100, and 1000 years.

landslides and tephra falls not only have lower probabilities than other hazards, but also would probably cause less damage, the former because of localization of damage to a small area, the latter because damage could probably be repaired at relatively low cost.

Although the discussion above emphasizes potential damage to existing buildings, roads, and bridges in Shakwak Valley, a relative ranking of natural hazards broadly similar to that shown in Figure 17 applies also to anticipated future developments. For example, floods, debris torrents, and earthquakes are the most dangerous hazards to planned as well as existing roads and structures in the short and intermediate term. However, for large expensive planned structures such as the gas pipeline, measures may be taken to mitigate against some hazards. For instance, adequate burial of a pipeline at stream crossings should protect it against flooding and channel and bank scour.

## CONCLUSIONS

In remote areas with inadequate historical records, natural hazards may be recognized and evaluated using a geomorphic approach. Terrain features produced by prehistoric geologic events, which today would be destructive to life and property, are often recognizable on air photos. These features and their associated deposits may be dated using various radiometric and relative age dating techniques, and the past frequency of potentially destructive events thus ascertained. This information, supplemented with available historical data, may be used to estimate the likelihood of future damaging events and to compare the relative dangers posed by the various hazards. Although this approach cannot be used for precise predictions as to the timing, magnitude, and exact location of destructive geologic events, it does provide a general first-order understanding of natural hazards and their interrelationships in a regional context.

A wide variety of natural hazards in Shakwak Valley have been recognized and evaluated using primarily a geomorphic investigative approach. It is concluded that most potential hazards are unlikely to cause major damage to existing towns and roads in the short term (10-year time frame); that floods, debris torrents, and earthquakes are the greatest hazards in the intermediate term (100-year time frame); and that inundation of Haines Junction by Lake Alsek, large landslides, and explosive volcanism have much lower probabilities and consequently should be of less concern.

## ACKNOWLEDGMENTS

This study is part of a research program concerned with the environmental aspects relating to pipeline concerns in southern Yukon Territory. This program is partially funded by Foothills Pipe Lines (South Yukon) Ltd. Logistical support for field operations was provided by R.B. Campbell (Geological Survey of Canada). V.N. Rampton (Terrain Analysis and Mapping Services Ltd.) and R.B. Horner (Earth Physics Branch) supplied information and advice on the Quaternary geology and historical seismicity of the area, respectively. W.H. Mathews (University of British Columbia) read and commented on a draft of the paper.

## REFERENCES

Berger, A.R. 1960. On a recent volcanic ash deposit, Yukon Territory. *Geol. Assoc. Canada Proc.* **12**, 117-18.

Bostock, H.S. 1948. *Physiography of the Canadian Cordillera, with special reference to the area north of the Fifty-Fifth Parallel.* Geol. Survey Canada Mem. 247, 106 p.

Bostock, H.S. 1952. *Geology of northwest Shakwak Valley, Yukon Territory.* Geol. Survey Canada Mem. 267, 54 p.

Boucher, G. and T.J. Fitch 1969. Microearthquake seismicity of the Denali Fault. *Jour. Geophys. Research* **74**, 6638-48.

Broscoe, A.J. and S. Thomson 1969. Observations on an alpine mudflow, Steele Creek, Yukon. *Canadian Jour. Earth Sci.* **6**, 219-29.

Brown, R.J.E. 1967. *Permafrost in Canada.* Geol. Survey Canada Map 1246A.

Campbell, R.B. and C.J. Dodds 1978. Operation Saint Elias, Yukon Territory. In *Current research, part A.* Geol. Survey Canada Paper 78-1A, 35-41.

Capps, S.R. 1916. An ancient volcanic eruption in the upper Yukon basin. In *Shorter contributions to general geology, 1915.* U.S. Geol. Survey Prof. Paper 95, 59-64.

Clague, J.J. 1979a. The Denali Fault System in southwest Yukon Territory – a geologic hazard? In *Current research, part A.* Geol. Survey Canada Paper 79-1A, 169-78.

Clague, J.J. 1979b. An assessment of some possible flood hazards in Shakwak Valley, Yukon Territory. In *Current research, part B.* Geol. Survey Canada Paper 79-1B, 63-70.

Clague, J.J. and W.H. Mathews 1973. The magnitude of jökulhlaups. *Jour. Glaciology* **12**, 501-4.

Collins, S.G. and G.K.C. Clarke 1977. History and bathymetry of a surge-dammed lake *Arctic* **30**, 217-24.

de Laguna, F. 1972. Under Mount Saint Elias: the history and culture of the Yakutat Tlingit. *Smithsonian Contr. Anthropology* **7** (3 pt.), 1395 p.

Denton, G.H. and M. Stuiver 1966. Neoglacial chronology, northeastern St. Elias Mountains, Canada. *Am. Jour. Sci.* **264**, 577-99.

Eisbacher, G.H. 1976. Sedimentology of the Dezadeash flysch and its implications for strike-slip faulting along the Denali Fault, Yukon Territory and Alaska. *Canadian Jour. Earth Sci.* **13**, 1495-513.

Eisbacher, G.H. and S.L. Hopkins 1977. Mid-Cenozoic paleogeomorphology and tectonic setting of the St. Elias Mountains, Yukon Territory. In *Report of activities, part B.* Geol. Survey Canada Paper 77-1B, 319-35.

Fahnestock, R.K. 1969. Morphology of the Slims River. *Icefield Ranges Research Proj. Sci. Results* **1**, 161-72.

Foothills Pipe Lines (South Yukon) Ltd. 1979. *Environmental impact statement for the Alaska Highway Gas Pipeline Project.* Calgary, Alberta: Foothills Pipe Lines (South Yukon) Ltd., 469 p.

Glaciology Division 1977. *Report on the influence of glaciers on the hydrology of streams affecting the proposed Alcan pipeline route.* Vancouver, B.C.: Canada Department of Fisheries and the Environment, Inland Waters Directorate, 38 p.

R.M. Hardy & Associates Ltd. 1978. *Report on laboratory testing on lake-bottom samples from Kluane Lake, Nisutlin Bay and Teslin River.* Calgary, Alberta: R.M. Hardy & Associates Ltd., 24 p., 4 app.

Hickman, R.G. and C. Craddock 1973. Lateral offsets along the Denali Fault, central Alaska Range, Alaska. *Geol. Soc. America Abs. with Programs* **5**, 322.

Hickman, R.G., C. Craddock and K.W. Sherwood 1977. Structural geology of the Nenana River segment of the Denali Fault System, central Alaska Range. *Geol. Soc. America Bull.* **88**, 1217-30.

Hughes, O.L., V.N. Rampton and N.W. Rutter 1972. *Quaternary geology and geomorphology, southern and central Yukon (northern Canada).* 24th Internat. Geol. Cong. Guidebook, Field Excursion All, 59 p.

Johnson, F. and H.M. Raup 1964. Investigations in southwest Yukon: geobotanical and archaeological reconnaissance. *Robert S. Peabody Found. Archaeol. Papers* **6**(1), 3-198.

Kindle, E.D. 1952. *Dezadeash map-area, Yukon Territory.* Geol. Survey Canada Mem. 268, 68 p.

Lanphere, M.A. 1978. Displacement history of the Denali Fault System, Alaska and Canada. *Canadian Jour. Earth Sci.* **15**, 817-22.

Lerbekmo, J.F. and F.A. Campbell 1969. Distribution, composition, and source of the White River ash, Yukon Territory. *Canadian Jour. Earth Sci.* **6**, 109-16.

Lerbekmo, J.F., J.A. Westgate, D.W.G. Smith and G.H. Denton 1975. New data on the character and history of the White River volcanic eruption, Alaska. In *Quaternary studies: selected papers from IX INQUA Congress,* R.P. Suggate and M.M. Cresswell (eds.). Royal Soc. New Zealand Bull. **13**, 203-9.

Mathews, W.H. 1965. Two self-dumping ice-dammed lakes in British Columbia. *Geog. Rev.* **55**, 46-52.

Mathews, W.H. 1973. Record of two jökullhlaups. In *Symposium on the hydrology of glaciers*. Internat. Assoc. Sci. Hydrology Pub. 95, 99–110.

McConnell, R.G. 1905. The Kluane mining district. In *Summary report on the operations of the Geological Survey for the year 1904*. Geol. Survey Canada Ann. Rept. (N.S.) 16, 1A–18A.

Muller, J.E. 1967. *Kluane Lake map-area, Yukon Territory (115G, 115F E½)*. Geol. Survey Canada Mem. 340, 137 p.

Plafker, G., T. Hudson and D.H. Richter 1977. Preliminary observations on late Cenozoic displacements along the Totschunda and Denali Fault Systems. In *The United States Geological Survey in Alaska: accomplishments during 1976*, K.M. Blean (ed.). U.S. Geol. Survey Circ. 751–B, B67–B69.

Plafker, G., T. Hudson, T. Bruns and M. Rubin 1978. Late Quaternary offsets along the Fairweather Fault and crustal plate interactions in southern Alaska. *Canadian Jour. Earth Sci.* 15, 805–16.

Post, A. and L.R. Mayo 1971. *Glacier dammed lakes and outburst floods in Alaska*. U.S. Geol. Survey Hydrol. Invest. Atlas HA–455, map and explanatory text.

Rampton, V.N. 1969. Pleistocene geology of the Snag-Klutlan area, southwestern Yukon, Canada. Unpublished Ph.D. thesis, Univ. Minnesota, Minneapolis, 237 p.

Rampton, V.N. 1970. Neoglacial fluctuations of the Natazhat and Klutlan Glaciers, Yukon Territory, Canada. *Canadian Jour. Earth Sci.* 7, 1236–63.

Rampton, V.N. 1979a. *Surficial geology and geomorphology, Mirror Creek, Yukon Territory*. Geol. Survey Canada Map 4–1978.

Rampton, V.N. 1979b. *Surficial geology and geomorphology, Koidern Mountain, Yukon Territory*. Geol. Survey Canada Map 5–1978.

Rampton, V.N. 1979c. *Surficial geology and geomorphology, Burwash Creek, Yukon Territory*. Geol. Survey Canada Map 6–1978.

Rampton, V.N. 1979d. *Surficial geology and geomorphology, Generc River, Yukon Territory*. Geol. Survey Canada Map 7–1978.

Rampton, V.N. 1979e. *Surficial geology and geomorphology, Congdon Creek, Yukon Territory*. Geol. Survey Canada Map 8–1978.

Rampton, V.N. 1981. *Surficial materials and landforms of Kluane National Park, Yukon Territory*. Geol. Survey Canada Paper 79–24, 37 p.

Read, P.B. and J.W.H. Monger 1976. *Geology and mineral deposits maps of Kluane and Alsek Ranges, Yukon Territory (parts 115A, B, G)*. Geol. Survey Canada, Open File 381, 6 maps.

Richter, D.H. and N.A. Matson, Jr. 1971. Quaternary faulting in the eastern Alaska Range. *Geol. Soc. America Bull.* 82, 1529–39.

Stevens, A.E. and W.G. Milne 1974. A study of seismic risk near pipeline corridors in northwestern Canada and eastern Alaska. *Canadian Jour. Earth Sci.* 11, 147–64.

Stone, K.H. 1963. The annual emptying of Lake George, Alaska. *Arctic* 16, 26–40.

Stout, J.H., J.B. Brady, F. Weber and R.A. Page 1973. Evidence for Quaternary movement on the McKinley strand of the Denali Fault in the Delta River area, Alaska. *Geol. Soc. America Bull.* 84, 939–47.

Stuiver, M., H.W. Borns, Jr. and G.H. Denton 1964. Age of a widespread layer of volcanic ash in the southwestern Yukon Territory. *Arctic* 17, 259–60.

Tempelman–Kluit, D.J. 1974. *Reconnaissance geology of Aishihik Lake, Snag and part of Stewart River map-areas, west-central Yukon (115A, 115F, 115G and 115C).* Geol. Survey Canada Paper 73-41, 97 p.

Terrain Analysis and Mapping Services Ltd. 1978. *Geology and limnology of Kluane Lake. I. Preliminary assessment.* Geol. Survey Canada Open File 527, 51 p.

# 3

# LANDFORMS FOR PLANNING USE IN PART OF PIERCE COUNTY, WASHINGTON

*Allen J. Fiksdal*

## ABSTRACT

In a geologic study for land use planning in part of western Washington State, landforms were mapped to define readily recognizable areas of similar geological constraints and were keyed to the distribution of (a) earth materials, (b) mineral resources, (c) engineering characteristics, and (d) geologic processes and hazards. Sixteen landforms were differentiated with distinctive characteristics which provide useful planning units, identifying the distribution of similar geologic materials, processes, and origin.

## INTRODUCTION

Pierce County, Washington, is an area of rapid population growth where expansion of industrial, commercial, and urban development threatens prime agricultural land (Fig. 1). Currently, the county government is developing a comprehensive land use plan that will conserve the rapidly dwindling agricultural land while providing for the needs of industrial, commercial, and residential growth. Geologic data, provided for the comprehensive land use plan, identified areas of natural constraints (hazards) and suitabilities (resources) for better utilization of the land. In Pierce County these constraints and suitabilities were identified and associated with specific landforms for ease of recognition and adaptability for land use planning purposes. Geologic maps are commonly used by planners to identify the natural conditions affecting land use planning, but these non Earth scientists usually do not have the expertise to identify possible hazards associated with specific geologic formations. In many cases, geologic maps have units grouped in time stratigraphic formations rather than lithologic units, which are the more useful for identifying areas with specific engineering conditions. Also, many geologic maps are "bedrock" maps, which do not delineate surficial units, except in large generalized groups (i.e., alluvium, glacial). Geologists have overcome some of these problems by making numerous interpretive or geologic factor maps that identify the hazardous zones, areas of natural resources, and surficial engineering properties for any given area.

In western Pierce County instead of developing the many interpretive or factor maps, a single map delineating landforms, and a table, listing the constraints and suitabilities for each landform, can be used more effectively. Because landforms are produced by specific geologic processes that over a period of time erode deposits, or shape the Earth's surface into a recognizable form, a landform map can delineate areas of similar topography and geologic conditions. The landforms can then be used as a basin land planning unit.

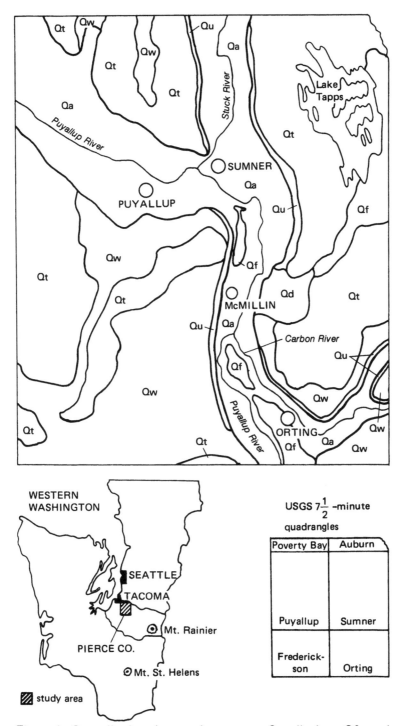

**Figure 1** Generalized geology and area map. Qa, alluvium; Qf, mud-flow; Qw, outwash; Qd, outwash delta; Qt, Lodgment till; Qu, glacial and nonglacial deposits undifferentiated.

This idea is not new; for example, McHarg (1969) used physiographic provinces to determine areas of similar landforms for planning purposes on the East Coast and Palmer (1976) and the Oregon Department of Geology and Mineral Industries have used landforms in their mapping of environmental geology along the Oregon coast. This study extends those by developing a landform classification system in glacial terrain for use in regional planning in western Washington State. No fixed precedent for the selection of planning landform units exists for this region, although Crandell (1963) did differentiate some landforms for geomorphic description in his work in part of Pierce County. This part of the Puget Lowland has been subjected to at least four advances of Cordilleran ice during the Pleistocene (Crandell et al. 1958), and eleven of the sixteen landforms identified reflect the occupation and withdrawal of the continental ice. The other five landforms are the result of recent (Holocene) erosion or depositional processes, including mudflows from Mount Rainier to the east.

The landforms were identified from 1:24000 scale air photography and geologic and topographic maps. The geologic maps differentiated the major groups (i.e., those underlain by till outwash or alluvium), and the surface morphology allowed for further landform differentiation (Table 1). With each landform, differentiated surface processes and hazards associated with each were identified. The engineering characteristics and mineral resources were identified for each landform from geologic maps and field reconnaissance. Thus the landform provided the base unit with which hazards, engineering characteristics, and mineral resources could all be associated.

The landforms were mapped at a scale of 1:24000 using U.S. Geological Survey 7½-minute topographic quadrangles as base maps. Over 500 km² were mapped using a

**Table 1**  Criteria for landform recognition

| Landform | Surface morphology | Lithology |
|---|---|---|
| Drumlin | Elongated ridges[a] | Till |
| Ground moraine | Lack integrated drainage and elongated ridges[a] | Till |
| Constructional slope | Lack elongated ridges, slope 12–40%[a] | Till |
| Outwash surface | Hummocky ground[a] | Sand and gravel |
| Outwash valley train | Abandoned outwash channels[a] | Sand and gravel |
| Kame terrace | Prominant terraces,[a] associated with outwash channels and collapse features | Sand and gravel |
| Terrace undifferentiated | Small terraces[a] at varying elevations | Sand and gravel |
| Collapse slope | Slump and flowage topography[a] | Sand and gravel |
| Outwash delta | Delta terrace at mouth of outwash channel[a] | Sand and gravel |
| Valley train | Holocene alluvial valleys[a] | Sand, gravel, silt, organic deposits |
| Erosional slope | Steep (40%) undercut slopes[a] | Variable |
| Mudflow surface | Flat surface | Mudflow debris[a] |
| Modified slope | Subtle, smooth hummocky surface | Lacustrine silts[a] |
| Kettles | Depressions in outwash areas[a] | |
| Esker | Sinuous ridge[a] | Sand and gravel |
| Modified area | Man-made cuts and fills[a] | |

[a]Major identifiable features.

composite base of all or parts of four quadrangles. Figures 2 and 3 are examples of small portions of the map.

## LANDFORMS

The landforms are grouped below for ease of discussion, based on their land use constraints and suitabilities. The first group is characterized by poor infiltration because of underlying till; the second group is basically outwash and alluvial deposits with good infiltration but also having areas where local flooding may occur. The landform derived from slope erosion and mass wasting makes up the third group, and the fourth group is comprised of local dissimilar landforms with varying characteristics.

### Drumlins, ground moraines, and constructional slope landforms

The drumlins, ground moraines, and constructional slopes are landforms underlain by till. Drumlins are found throughout the area but are most obvious in the northeast quarter, where they form elongated islands in Lake Tapps. The drumlins are not "classical" drumlins, with a spoon-shaped form, but instead are long, symmetrical linear features composed of till. The drumlins vary in size from 3 to 33 m high and 0·3 to 2 km long. They are aligned north–south, reflecting the direction of movement of the continental ice. Ground moraine is defined by Flint (1957 p. 131) as a "moraine having low relief

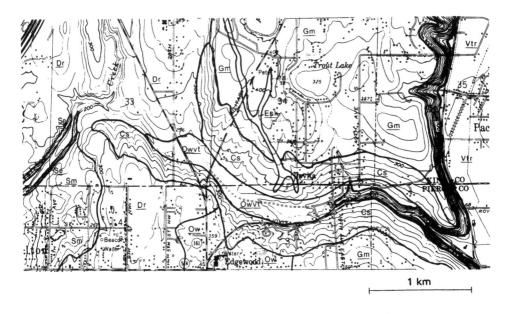

**Figure 2** Section of landform map with landforms outlined. Gm, ground moraine; Sm, modified slope; Ow, outwash; Ma, modified area; Dr, drumlin; Se, erosional slope; Owvt, outwash valley train; Cs, constructional slope; Vtr, valley train; Esk, esker; ——, contact. (See text for detailed explanation of map units.)

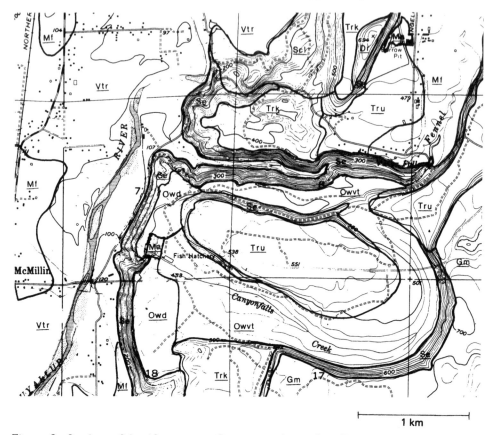

**Figure 3**  Section of landform map. Grm, ground moraine; Dr, drumlin; Tru, terrace undifferentiated; Mf, mudflow; Ma, modified area; Vtr, valley train; Trk, kame terrace; Se, erosional slope; Owvt, outwash valley train; Owd, outwash delta; Scl, collapse slope; ——, contact. (See text for detailed explanation of map units.)

devoid of transverse linear elements." These areas are underlain by till, do not have drumloidal features, lack integrated drainage, and have a slope less than 12%. They contain closed depressions and are usually adjacent to drumloidal landforms. Constructional slopes are landforms underlain by till, are devoid of transverse linear elements, and have slopes of approximately 12 to 40%.

The lodgment till that underlies these landforms is an unsorted, compacted mixture of silt, sand, and gravel formed beneath the overriding ice sheet. Because of its mixture of particle sizes and its compaction by the ice, till has low permeability. This inhibits surface water infiltration on all three of the landforms noted above. Poor infiltration causes problems primarily for septic sewer systems, which rely on effluent percolation, ponding, and runoff. Such septic systems cannot function properly because of the till's low permeability. During the wet winter months in areas of poor drainage and closed depressions on the ground moraine landform, surface water collects in the depressions and causes local flooding and failure of septic systems. Contamination of perched aquifers also occurs in these areas because of their proximity to the ground surface. Runoff of surface water is not a significant problem on the drumlins because of the short lengths of

| LANDFORMS | Percent of total land area | Landslides | Ponding | Gully erosion | Runoff | Settlement | Creep | Flooding | Groundwater contamination | Slope | Foundation stability | Infiltration | Slope stability | Sand | Gravel | Top soil | Fill |
|---|---|---|---|---|---|---|---|---|---|---|---|---|---|---|---|---|---|
| DRUMLIN | 25 | | | ○ | ○ | | ○ | | | | ○ | ● | ○ | | | | ○ |
| GROUND MORAINE | 5 | | ◑ | | | | | | | | ○ | ● | ○ | | | | ○ |
| CONSTRUCTIONAL SLOPE | 5 | ○ | | ○ | ○ | | ○ | | | | ○ | ● | ○ | | | | ◑ |
| OUTWASH SURFACE | 5 | | | | | | | | ◑ | | ○ | ○ | ○ | ○ | ○ | | ○ |
| OUTWASH VALLEY TRAIN | 10 | | ◑ | | | | | | ◑ | | ○ | ⊕ | ○ | ○ | ○ | | ○ |
| KAME TERRACE | 5 | | ○ | | | | | | ◑ | | ○ | ○ | ○ | ◑ | ◑ | | ○ |
| TERRACE UNDIFFERENTIATED | 2 | | | | | | | | | | ○ | ○ | ○ | ◑ | ◑ | | ○ |
| COLLAPSE SLOPE | 2 | ○ | | ○ | | | ○ | | | | ○ | ⊕ | ◑ | ◑ | ◑ | | ○ |
| OUTWASH DELTA | 1 | | | | | | | | | ● | ○ | ○ | ● | ○ | ○ | | ○ |
| VALLEY TRAIN | 20 | | ◑ | | ◑ | | ● | ◑ | | | ⊕ | ◑ | | ◑ | ◑ | ○ | ○ |
| EROSIONAL SLOPE | 10 | ● | | ● | ◑ | | ● | | | ● | ◑ | | ● | | ◑ | | ○ |
| MUDFLOW SURFACE | 5 | | ◑ | | | | | | | | ○ | ● | ○ | | | | ◑ |
| MODIFIED SLOPE | 1 | | ◑ | | | | ◑ | | | | ○ | ● | ○ | | | | ● |
| KETTLES | 1 | | ○ | | | | | | | | | ● | | | | | |
| ESKER | 1 | | | | | | | | | | ○ | ○ | ○ | ◑ | ◑ | | ○ |
| MODIFIED AREA | 3 | | ○ | | ◑ | | ○ | | | | ⊕ | ⊕ | ⊕ | | | | ● |

no symbol indicates non-applicability, no hazard, or no resource

○ slight    ○ good    ○ good
◑ moderate    ◑ moderate    ◑ moderate
● severe    ● poor    ● poor
⊕ variable

**Figure 4** Landforms and associated hazards, engineering characteristics, and mineral resources.

49

the slopes. It is, however, a moderate hazard on constructional slopes. Effluent runoff, in the form of springs, will sometimes cause problems on constructional slopes because of the poor infiltration.

Because of the precompaction of the till by the ice and lack of steep slopes, slope stability problems are minimal on these three landforms and foundation stability should be excellent, even during earthquakes and other ground-shaking events (Crandell 1963). These landforms are not a significant source of mineral resources. Till is too unsorted to be used as an aggregate source, so it is used only as fill material. Peat is found in a few of the larger closed depressions but is of minor economic value.

Thus the major planning significance of the drumlins, ground moraine, and constructional slopes is poor infiltration capabilities, local ponding, and lack of resource (gravel). These landforms do, however, provide areas of excellent foundation stability and have minimal slope stability problems (Fig. 4).

## Outwash surface, outwash valley train, kame terrace, collapse slope, terrace undifferentiated, outwash delta, and valley train landforms

This group of landforms is characterized by being underlain predominately by sand and gravel deposited by fluvial processes. The valley train differs from the others by having overbank fine-grained flood deposits and organic accumulations.

The outwash surface and outwash valley train occur in uplands throughout the area but are most notable south of Puyallup and Tacoma (see Fig. 1). These landforms are the result of fluvial deposition and erosion from melt waters of the continental glacier. Outwash surfaces (distinguished by hummocky gravelly terrain) are areas where sand and gravel was deposited on a till-covered upland. The water and septic effluent percolate very well. However, groundwater contamination is a dangerous possibility because there is no confining layer (like till) protecting underlying aquifers. Thus, groundwater protection must be considered. In smaller outwash valley trains ponding and septic system failure occur in some areas where silt and organic deposits have accumulated, or where glacial meltwaters have scoured a channel in till. These areas are usually small, extending for only 0·5 to 2 km along the outwash channels.

The stratified drift found in this group of landforms provides good foundation stability during earthquakes, and slope stability problems are minimal. However, because the collapse slopes have lost their internal layered structure, and may have slopes as steep as 60%, they could become unstable if significantly altered. Undercutting and overloading slopes, redirection of drainage, and addition of groundwater may decrease stability. The outwash delta has slopes at the angle of repose that are actively eroding.

The outwash valley train is an excellent source of aggregate, and many sand and gravel mining sites are located here. This area is important as a mineral resource area. Aggregate on the kame and undifferentiated terraces has also been mined, but it often has a high percentage of sand.

The valley train landform not only contains stratified outwash sediments but overbank flood silts, peat, and organic soils as well. It has special planning significance because of these materials, their engineering characteristics, and the hazards associated with active processes on this landform.

The large Puyallup River valley train was formed by river and glacial erosion and deposition. The valleys are underlain by fluvial and marine sediments and recent alluvium.

Parts of the alluvium south of Sumner are overlain by mudflow deposits and will be described later under mudflow surfaces. The alluvium in the valley train ranges in particle size from fine to coarse, saturated to nonsaturated, and may have a low to high organic content. Because of overbank flooding and river meandering, the alluvium is not a uniform layering of sediments but consists predominately of lenses and pockets of varying-size material.

In many areas of the valley train landform, the groundwater table is at, or very near, the surface, resulting in saturation of the alluvium. When saturation occurs, septic systems fail and effluent may flow to the surface or contaminate perched aquifers. Also, saturation of the fine-grained sediments may result in differential settlement by realignment of individual grains and reduction of pore space and loss of mass soil volume, with the potential for damage to building foundation. During an earthquake, differential settlement is also likely to occur in the saturated sediments. Crandell (1963) reviewed the effects of three major earthquakes that were felt in the project area in 1939, 1946, and 1949. During these events, structures in the Puyallup Valley on recent alluvium were damaged more severely than those in upland areas. Peat deposits also have poor foundation and seismic stability; they may settle excessively due to the high water content and compressibility.

Flooding is another important hazard in the valley trains; however, most major flooding is controlled by dikes along the rivers. Local seasonal flooding does occur in low saturated depressions and where side streams are prevented from draining by flood control dikes.

The major resource of the valley train is the fertile soils developed on the alluvium. These valley trains are the prime agricultural area in Pierce County. Another resource is the sand and gravel deposited by the rivers, but because of the proximity to large areas of well-sorted sand and gravel (the upland outwash areas), problems of river siltation, contamination, change in river dynamics, and loss of agricultural land, it is not recommended as an aggregate resource area.

The major planning concerns of the valley train are the optimum utilization of the agricultural resource, potential for structural damage during an earthquake, contamination of shallow groundwater, and flooding.

## Erosional slope landform

The erosional slope landform is found in the areas that have undergone or are still undergoing erosion by stream undercutting and mass wasting. These slopes are the steep (greater than 70%) hillsides of the Puyallup, Stuck, and Carbon Valleys. The erosional slopes are usually mantled by a few meters of colluvium, which covers the horizontally layered sequence of unconsolidated pre-Wisconsin lacustrine silt, mudflows from Mount Rainier, and glacial and nonglacial sands and gravels.

Almost all of the landsliding, mass wasting, and gully and stream erosion occur on the erosional slope landform. Change of these hillsides by excavations and artificial fill could cause increased instability by altering groundwater, runoff, angle of slope, and by the addition of weight. These steep hillsides are also subject to failure during earthquakes. Thus erosional slopes are the least desirable for land development. Specific site evaluations should be made for slopes over 50% and in areas of concentrated surface and groundwater.

## Mudflow surface, modified slope, kettles, esker, and modified area landforms

The landforms described in this section are all dissimilar to each other and usually have only local significance.

The mudflow surfaces are underlain by deposits from the Osceola and Electron Mudflows. The Osceola Mudflow flowed from the north side of Mount Rainier across the uplands east of Sumner about 5000 years ago. The Electron Mudflow came from the south side of Mount Rainier down the Puyallup River valley to about Sumner only 500 years ago. The mudflows are a mixture of silt, sand, and rock (similar to till but uncompacted) and have little resource value other than fill. In some cases, however, they may have a high clay content resulting in a need for control of moisture content when used as fill.

The mudflow, because of its similarity to till, has essentially the same characteristics as ground moraine. The materials have good foundation stability and poor infiltration. Bearing strength of the Osceola Mudflow is adequate for light construction in areas of high water tables because of an oxidized zone at the surface. Oxides have cemented 3 to 3·7 m of surface of the mudflow. Crandell (1963) observed a low bearing strength below the oxidized zone, where the mudflow becomes unstable if it is disturbed or reaches its liquid limit. Because of the oxidized zone, seismic stability is also good. In the Puyallup Valley where the Electron Mudflow overlies the river alluvium, the differential settlement effect of saturated alluvium is reduced during earth-shaking events because of the till-like, well-mixed clay, sand, and gravel that comprise the flow.

Mudflows from Mount Rainier are a serious but rare hazard event in the Puyallup Valley area. However, consideration of this hazard should be made before development takes place, especially in valley areas.

The modified slope landforms are hillsides mantled with recessional lake silts and clays. These lacustrine sediments usually overlie till and are limited in area, occurring near Milton and South of Puyallup. This landform is recognized primarily by its lithologic nature, but subtle smooth hummocky surface unlike other areas underlain by till can also be seen on air photographs. Their planning significance is their very poor permeability and lack of resource value. This landform does not have stability problems because slopes generally range only from 12 to 20%.

Kettles are found on portions of kame terraces near Orting. They were formed after buried blocks of stagnant glacial ice melted, causing collapse of the overburden. The kettles are as large as 30 m deep and 150 m across. In some kettles, silt and organic debris have accumulated and inhibited infiltration. Ponding occurs in these during wet winter months.

Another surficial feature left by the Vashon glacier is an esker. It was deposited by a stream flowing within, or beneath, the glacier and is now a topographic expression of that ancient stream course. This is the only esker found in the southern Puget Lowland and is therefore geologically unique. The esker is not important as an aggregate source because of its small size.

The most recent changes in landform have been those caused by man. Such landforms are defined here as "modified areas." They include major engineering earthwork structures and artificial cuts. Examples are the dikes along the rivers placed to restrict river migration, artificial fill that has been used to fill low areas (particularly at the mouth of the Puyallup River in the Tacoma industrial area), construction of roads and highways, and gravel mining and rock quarrying that have created large cuts and/or depressions in slopes and flat areas. In all, 2 to 3% of the total area studied has been modified.

Fill may be poorly controlled and have much organic debris, large boulders, clay, and other waste material. Because of this, it has no resource value and its engineering characteristics are extremely varied. These man-made features change the natural form of the land by adding or subtracting materials and changing natural infiltration patterns, stability, and strength characteristics. Unknown and highly variable conditions, materials, compaction, and moisture content point out the need for individual site investigations before development of uncontrolled fill areas. Artificial slopes need to be evaluated by site-specific studies because of unknown drainage, erosion potential, and possible loss of slope support through previous alteration of the natural conditions by oversteepening or overloading slopes.

## CONCLUSIONS

This study developed a landform classification system in glaciated terrain for use in regional land use planning, which proved to be a good form of delineating areas of similar suitabilities and constraints. Not only is the distribution of earth materials outlined (for distinguishing areas of specific engineering properties) but with landform mapping, areas of similar surface morphology for identifying hazards (i.e., ponding, mass wasting, flooding, etc.) can also be delineated.

Sixteen landforms were mapped and the hazards, engineering properties, and potential for resources were identified. The major planning significance of drumlins, ground moraine, and constructional slopes included their poor infiltration capabilities, local ponding on the ground moraine, and lack of resource; they also have excellent foundation stability and minimal slope stability problems. The landforms underlain by outwash deposits (outwash surface, outwash valley train, kame terrace, and collapse slope) all have excellent infiltration capabilities, little settlement problems, and good foundation stability (but there is the potential for groundwater contamination because of the excellent infiltration capabilities, and the collapse slope could possibly have slope stability problems). These landforms are underlain by sand and gravel and most are good resource areas. The best resource potential in this part of Pierce County is the outwash delta because of its quantity and quality of sand and gravel.

The valley train has great planning significance because of its prime soil resource, its potential for flooding, differential settlement, high water table, and others. It is also the area under most pressure for development. Another important landform is the erosional slope. This landform is the result of recent erosional processes and is the location of most of the mass wasting occurring in this area. Slopes are up to 100% and any undercutting or overloading of these slopes will probably result in landsliding. This area is very important to any land use plan because of its mass-wasting hazard.

The mudflow surface, terrace undifferentiated, modified slope, kettles, esker, and modified area landforms are all rather limited in area and have variable constraints and suitabilities. Mudflow surfaces have poor infiltration as do modified slopes and the bottom of kettles. The esker is the only one found in the lower Puget Lowland and therefore geologically unique and should possibly be designated as a unique area. Areas modified by man are usually unpredictable and need careful attention when determining their constraints and suitabilities.

## REFERENCES

Crandell, D.R. 1963. *Surficial geology and geomorphology of the Lake Tapps quadrangle, Washington.* U.S. Geol. Survey Prof. Paper 388-A.

Crandell, D.R., D.R. Mullineaux, and H.H. Waldron, 1958. Pleistocene sequence in southeastern part of the Puget Lowland, Washington. *Am. Jour. Sci.* **256**, 384-97.

Flint, R.J. 1957. *Glacial and Pleistocene geology.* New York: John Wiley. 553 p.

McHarg, I.L. 1969. *Design with nature.* New York: Natural History Press, 197 p.

Palmer, L. 1976. *Application of land use constraints in Oregon.* Geol. Soc. America Spec. Paper 174, 61-84.

# 4

# GEOMORPHOLOGY AS AN AID TO HAZARDOUS WASTE FACILITY SITING, NORTHEAST UNITED STATES

*Allen W. Hatheway and Zenas F. Bliss*

## ABSTRACT

Nearly every conceivable option for management of hazardous waste entails some form of interaction with the ground, whether in processing waste toward neutralization or in the secure land burial of the waste or its residue of processing.

Almost all potential hazardous waste sites in the Northeast involve some aspect of hydrogeological or geological impact. Geological influences can profoundly affect the cost of development of a disposal or processing facility in terms of foundation design requirements. Some public officials have argued that geological influences are of low priority and the main siting concern should be in the sociopolitical aspects of implementation; disregard of geological factors in siting could be disastrous in terms of additional construction costs or adverse environmental impact. Geologic assessments are a must at every level of site selection and should be implemented at the earliest possible time in the site qualification process.

Geomorphic indicators provide the entry point in determining what siting options are actually available in a given state or county area. The authors prefer a version of the surficial engineering geologic mapping scheme of Galster (1977) and employ this to produce areal maps depicting geomorphic units of similar engineering and hydrogeologic properties.

This approach to regional and site-specific mapping also produces map units which portray the physical properties and geologic conditions which are most important in meeting state and federal hazardous waste management facility siting guidelines (US EPA 1980).

The region is mapped on a data compilation basis from existing sources; potentially attractive sites are identified and rated for environmental engineers, who then choose one or more candidate sites. The most attractive site is then field mapped according to the geomorphic scheme and a site exploration plan is developed to prove the site and to determine design-related physical properties, the groundwater assessment, and three-dimensional extent of the surficial geologic units.

## HAZARDOUS WASTE AND ITS MANAGEMENT

The industrialized Northeast produces a formidable amount of hazardous wastes annually. Although strictly accurate accounts of this volume are not yet available, it is likely that the seven upper Northeast states produce an aggregate total of more than

55

1 900 000 000 liters of such wastes annually. At the present time liquid volume equivalents are generally used as the measure of volume because about 90% of hazardous wastes are in liquid form and the remainder is generally transported in barrels or tank trucks.

Hazardous wastes are defined and governed by Section 3004 of the Federal Resource Conservation and Recovery Act (RCRA) of 1976, as administered by the U.S. Environmental Protection Agency (USEPA). The wastes are classified as being: toxic, corrosive, reactive, flammable, explosive, biologically viral, or of a low-level radioactivity. It is important to note that electric utility generation station fossil fuel wastes are not regarded as hazardous waste. The intent of RCRA is to see that the wastes are registered or manifested into the regulatory system upon generation and that they are eventually treated or permanently isolated to avoid possible adverse effects on humans, livestock, or the environment in general. RCRA mandates state control over this process and the states have been given the opportunity to devise their own management facility siting criteria, as long as these requirements meet or exceed the intent of the federal criteria. The federal criteria have been released for public, industry, and government comment since December 1978 and have not yet been finalized.

There is little consensus among the many concerned parties as to how the RCRA or applicable state objectives should be met. At the time of this writing there appears to be a broadly based contention that most of the waste should be processed in a variety of ways. Hence we use the overall term *management* to describe the collection, treatment, and disposal of hazardous wastes.

Collection and treatment of hazardous waste is generally outside the expertise of geologists, but the *management* concept has important implications for geologists. For many interested parties the concept of management connotes total process conversion to usable by-products, chemically inert substances, or energy fuels and feedstocks. Although this is indeed an admirable goal, it is our opinion that some form of residue will remain from the management process and that for these wastes, secure land burial will probably be the only acceptable manner of disposal. As geologists, it will be our responsibility to work toward siting and design of processing and disposal facilities which will be built at the least possible cost to the public and industry and which will have the most favorable associated environmental risks.

## GEOLOGY APPLIED TO FACILITY SITING

Hazardous waste management facilities must be designed so that processing activities will not compromise the environment due to inadvertent or accidental leaks or spills, and residue disposal sites must also be designed for total containment.

This containment objective can be met by large outlays of funds in constructing engineered retention features to include artificial liners, various barriers, and isolating schemes. However, in light of present geological siting capabilities, this does not constitute good engineering and so could waste public and private development funds that could be better used for societal needs or in combatting inflation. Accordingly, it will be most prudent to carefully select sites that possess the best possible mix of highly favorable geologic characteristics relating to material properties and hydrogeological conditions.

RCRA facility design criteria have two major hydrogeologic objectives: prevention of groundwater intrusion into the waste or processing residue and prevention of leachate migration from the containment area. These objectives are met by siting and constructing the facility in such a way as to avoid potential contact with groundwater and to utilize natural materials possessing low permeability.

Subsurface geological and geophysical exploration is a costly undertaking and is best employed after key siting decisions have been made and a candidate site or sites have been selected. There remains one powerful geological technique that can be used very effectively in the site selection process: geomorphological analysis.

## BROAD-AREA SEARCHES

If the owner of the proposed hazardous waste management facility has a preselected site or a limited number of candidates, preliminary site qualifying work can begin at once. As in the case of most engineering geological investigations, such studies are conducted at progressively more detailed levels until the site is either disqualified or designed.

In the case in which relatively large volumes of waste are to be processed and disposed of, or for those facilities designed to serve relatively large areas of waste generation, it is often more productive to weigh all important siting factors so that a least impact and least costly (to construct) site can be located.

We have termed such investigations *broad-area searches*, relying on available geologic and hydrogeologic maps and soil survey reports for preliminary site identification data rather than on actual field work.

The basic concept of the broad-area search is to develop a surficial geologic map of the area of interest and to portray on this map such geologic units as can be identified as having predictable hazardous waste management facility siting characteristics. For the Northeast, the units have been identified in the following section and are identified by their typical geomorphic form (Fig. 1).

Once the geomorphic map units have been defined, all available geologic and soil survey map references are interpreted to produce a siting-related surficial geologic map, an example of which is shown as Figure 2a. This example was developed as a study of methods of geological interpretation of multicounty areas as a basis for screening portions of entire states or the states themselves, without resort to costly and time-consuming field visits and by avoiding an initial biasing that would tend to originate without the benefit of comparing various options relating to geologic character and topographic conditions.

From the surficial geologic map, and in discussions with project environmental engineers, a basic concept of facility design is developed, and the surficial geologic map units are ranked according to generalized facility siting suitability (Fig. 2b). In this example, two general levels of suitability are presented for hazardous waste management facility siting according to the siting concepts for a particular project. When the management concept has progressed further and facility design characteristics are better known, the suitability factoring can be used to provide more specific differentiation of latitutde in surficial geologic map units affecting siting (Fig. 2c).

If the broad-area search surficial geologic map covers a relatively large area, say several counties, a variety of siting-related geologic and geotechnical engineering data can be compiled in matrix form and compared for relative differences between surficial geologic units (Table 1).

A siting matrix should be made up of physical property data and other character-

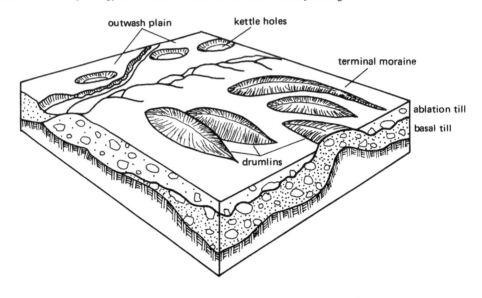

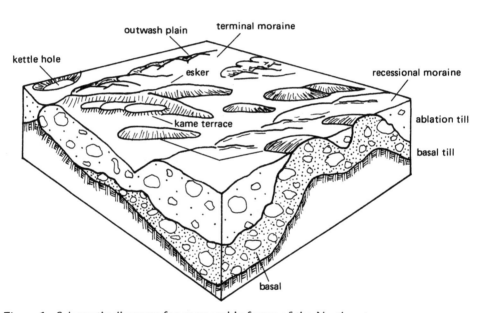

**Figure 1**    Schematic diagrams for geomorphic forms of the Northeast.

istics which are felt to control the utility of given surficial geologic units for hazardous waste management facility siting. Since the units are to be compared with each other within discrete areas (say counties or other jurisdictional units), the matrix data will probably contain many type entries which are familiar to most geologists. Of those factors shown on Table 1, perhaps only the *soil unit suitability factor* will be unfamiliar to the reader. This factor was developed by geotechnical engineers at Cornell University (Roberts & Sangrey 1977) as a representation of the relative ability of given SCS agronomic soil units to attenuate or diminish the concentration of an average groundwater-

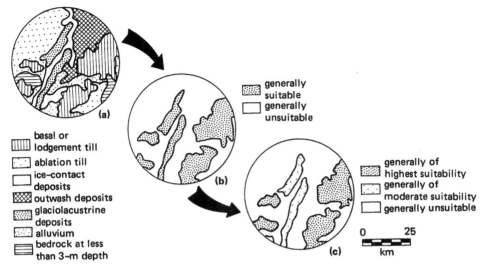

basal or lodgement till

ablation till

ice-contact deposits

outwash deposits

glaciolacustrine deposits

alluvium

bedrock at less than 3-m depth

generally suitable

generally unsuitable

generally of highest suitability

generally of moderate suitability

generally unsuitable

0    25

km

**Figure 2** Surficial geologic map and two derivative maps used to portray the areal distribution of geomorphically influenced map units which were selected as having important and predictable hazardous waste management facility siting characteristics. (a) the base map, showing six surficial map units in an area of western New York. This map was produced as an interpretation of the Niagara Sheet, Surficial Geologic Map of New York, by Ernest H. Muller, 1975; New York State Geological Survey. (b) Areal distribution of lumped geomorphic units, showing generalized siting suitabilities. (c) Final determination, according to the objectives of one particular siting study, of areas that are suitable for finalized siting studies relating to demography and transportation costs. Individual target areas would probably consist of 5 to 20 square miles and further determinations regarding suitability would require photogeologic interpretation, field mapping, subsurface explorations, and laboratory testing of soil samples.

**Table 1** Surficial geologic data types: siting matrix

The following data types were collected in the course of a recent comprehensive hazardous waste management study for the New York Environmental Facilities Commission. The data were collected or interpreted from existing references.

*County* for which the data apply.

*Name of the surficial geologic unit:* only those units which are generally present in the county or which have been mapped geologically, or which can be inferred from USDA soil survey data were included.

*Dominant lithology* of the soil units: refers to the parent rock material which is believed to have contributed the most volume to the soil unit; symbols are as follows:

| | | | |
|---|---|---|---|
| Ls | limestone | Cg | conglomorate |
| Ss | sandstone | Gr | Granitic (igneous rock) |
| Slt | siltstone | Gn | Gneissic (metamorphic rock) |
| Sh | shale | I | undifferentiated igneous |

*Typical thickness:* best estimate of the average thickness or range of thickness for this unit in the subject county. This presumes that this unit is the dominant surficial soil unit in the map area shown on the working surficial geologic map

**Table 1**  Surficial geologic data types:  siting matrix (*continued*)

*Soil unit suitability factor:*  interpreted assessment of which of the five categories (I through V) of leachate attenuation potential (after Roberts & Sangrey 1977) apply to the soil unit considered. The data are not always sufficient to warrant this estimate.

*Hydrogen ion activity (pH):*  most dominant condition assigned on the basis of data presented in SCS county soil survey reports; where shown in pairs (e.g., B/A), the dominant condition is shown first; B = basic (greater than pH 6·5) and A = acidic (less than pH 6·5).

*Percent clay and silt-size materials:*  an estimate of average content of the map unit, for materials passing the (−) 200 standard sieve. This value is a useful factor in estimation of the in situ permeability of the soil unit, as well as its leachate attentuative capacity.

*Relative density:*  estimate of the degree to which individual particles have been compacted together during the geologic history of the soil unit. The data are useful in interpreting both foundation-bearing capacity for hazardous waste management facilities and in estimating in situ permeability. Density is estimated by experience of the engineering geologist, on the basis of geomorphic landform type and ease or difficulty of backhoe excavation. Verification comes through the use of Standard Penetration Testing (SPT) during drilling, in which the split-spoon or other soil sampling device is driven into the soil by a known amount of foot-pounds of energy. The number of blows required ($N$) is generally correlatable to density, bearing capacity, and settlement characteristics to the soil. Laboratory density tests (in pounds per cubic foot or kilograms per cubic meter) are performed routinely in the laboratory on soil samples so obtained.

*Other geologic influences:*  naturally occurring geologic conditions which may influence hazardous waste management activities.

*Typical aquifer yield:*  best estimate of the typical yield of groundwater from pumped wells placed in the particular geologic unit.

*Nature of aquifer producing zone:*  assessment of the material making up the specific soil unit, regardless of its ability to contain water under conditions commonly regarded as indicating an aquifer. The assessment is generally made by reference to type of material and with respect to available well data in the area of investigation. For purposes of estimation, the engineering geologist and hydrogeologist think in terms of the coefficient of permeability (in centimeters per second of potential one-dimensional fluid flow) and in sustainable specific yield in gallons or liters per minute. Aquifers are generally avoided, when detected, in the siting process.

*Quality of water found in the unit:*  estimated in terms of federal drinking water standards.

*Suitability for construction of hazardous waste management facilities:*  overall assessment of the bearing capacity, permeability, leachate attenuation, and groundwater production characteristics in terms of placing a hazardous waste management facility in or on the particular soil unit. As in the case of physical property determinations made during the siting process, these suitability factors are generally rated as *highly suitable, suitable, marginally suitable*, or *unsuitable*. A strict determination of each factor must be made as soon as a site or candidate site is selected. The determinations are made on the basis of the areal and volumetric distribution of each of the geologic units that will be affected by or will affect the facility. Factors that are determined to be less than *highly suitable* will probably require some form of engineering in compensation for their deficiencies. The ratings are relative and apply only within the specific county.

borne waste leachate. The attenuation potential for a given soil type is largely a measure of the presence of clay minerals and clay and silt-size particles in which a relatively large amount of surface area is present on individual soil particles. Generally speaking, the finer-grained soils are more attenuative and are hence more valuable in hazardous waste facility siting efforts.

Often the most critical aspect of siting is the identification of groundwater conditions and the prediction of seasonal and high-precipitation hydrogeologic variations as they will affect suitability and future operation of the site. Boreholes and observation wells must be placed strategically, on the basis of site geologic and geomorphologic interpretations, in order to define groundwater levels, quality, and flow directions. Often a groundwater flow net is superimposed on the site geologic map.

## GEOMORPHOLOGY, PHYSICAL PROPERTIES, AND GROUNDWATER CONDITIONS

The Northeast states of New England and New York are characterized by a glacially produced landscape of young and distinct landforms. These landforms are formed in surficial geological units known to engineering geologists broadly as *soils*, due to their unlithified nature. Just as each of the characteristic landforms was formed or sculpted by recent sedimentary and glacial processes, so too their resulting landforms have been influenced by the intrinsic engineering properties of each soil unit. These are the characteristics that should be sought in order to meet RCRA objectives (Table 2).

On the basis of their geologic characteristics affecting hazardous waste management facility siting, glacial and surficial soil units of the Northeast fall into two broad categories: cohesionless and cohesive, in the terminology of geotechnical engineering (Table 3).

Generally speaking, it is the cohesive soil classification that more nearly meets the RCRA guidelines (Table 2) for natural materials, either in situ or as engineered fill. The use of geomorphic indicators, as advocated by the authors, seeks to identify soil units favorable to hazardous waste management facility siting and to do so without resort to costly subsurface exploration activities in the site selection process.

Classification of surficial geologic units in hazardous waste management facility siting should pay particular attention to mode of origin as it affects hydrogeologic character. Most of the surficial geologic units of the Northeast are in some way related to glacial processes and suffer from a general imprecision in identification. As noted in Tables 2 and 3, it is desirable in hazardous waste siting to know basically if the particular

**Table 2** Geologic characteristics used to meet RCRA guidelines[a]

| Characteristic | RCRA guideline |
|---|---|
| High unit weight<br>Clay minerals present<br>Silt and clay-size<br>  materials | Permeability coefficient ($k$)<br>less than $10^{-7}$ cm/s |
| Relatively high content<br>  of silt and clay-size<br>  materials | Plasticity index $> 15$ |
| High unit weight<br>Fine-grained materials<br>Relatively thick unit of<br>  deposition | 1·5-m separation from<br>  groundwater |

[a] As found in RCRA draft guidelines of December 18, 1978, USEPA.

**Table 3**    Engineering classification of surficial soil units

| Soil unit | Classification |
|---|---|
| Lodgement (basal) till (Figs. 3–5) | Cohesive |
| Ablation till (Fig. 6) | Cohesive to cohesionless |
| Ice contact deposits | Cohesionless |
| Outwash (Fig. 7) | Cohesionless |
| Glaciomarine deposits | Cohesive |
| Glaciolacustrine deposits (Fig. 8) | Cohesive |
| Beach and eolian deposits | Cohesionless |
| Organic deposits | Cohesive |

surficial geologic unit is cohesive and whether it has a relatively high bulk density. Both factors are prime determinants relating to hydrogeological conditions in the particular geologic unit. The Northeast is also a region in which the terminology of surficial geology is clouded with imprecision through the use of such terms as *drift, till,* and *ground moraine,* in which strict nature of deposition is not imparted with the name. For purposes of engineering geology in general and for use in hazardous waste facility siting in particular, the authors take pains to employ terms which are as specific as possible and which connote the presence or absence of favorable hazardous waste siting conditions. This is not to say, however, that other important factors, such as topography, will not control the ultimate value of a given surficial geologic unit. The following is offered as the least ambiguous terminology available for siting purposes:

1. *Lodgement till. Lodgement,* or *basal till* (Figs. 3 to 5), is heterogeneous and well-graded (in the engineering sense) glacial debris which was deposited and compressed beneath the mass of Pleistocene ice sheets. Lodgement till is characteristically hard and dense and is composed of varying amounts of coarse fragments in a matrix of fine-grained

**Figure 3**    New England basal till exposed in a foundation excavation and depicting its generally high-density nature, along with a fine-grained and often cohesive silt and clay-sized matrix.

**Figure 4**  Lodgement till of an extremely dense and low-permeability variety.

materials, often exceeding 40 to 50% clay and silt-size fractions. Whatever clay minerals that are present provide a degree of leachate attenuation, or retention of waste liquids which may tend to leak from landfills. Hydrologeologically, lodgement till is of low permeability, and thus is not an important potential source of groundwater. It is a medium through which leachate might move at acceptable rates or not at all.

**Figure 5**  Fine-grained basal till characteristic of the Great Lakes region shown here with tensile fracturing developed with desiccation and open-faced stressing at a lake shoreline. As a land burial medium, the till is of very low permeability and offers some degree of leachate attenuation (25-cent piece for scale).

2. *Ablation till.* Although also comprised of glacial debris, ablation till (Fig. 6) is an entirely different material in terms of hazardous waste management facility siting. Ablation till represents a heterogeneous mix of debris carried on or in the ice sheet and deposited whenever the ice sheet was melting, wholly or partially. This material is generally less dense than lodgement till and contains much less silt and clay-size grains. Subsequently, ablation till nearly always has a higher degree of in situ permeability and therefore may represent a limited source of groundwater in a site area. The leachate attenuation ability of this material is usually considerably less than that of the lodgement till.

3. *Outwash.* Outwash deposits material are formed from the transport of sediments in glacial meltwater streams. It is generally poorly graded (in the engineering sense; see Fig. 7) and contains particles having a variety of grain sizes, usually sand size and larger. Outwash generally exhibits some rudimentary bedding with grain size distribution varying more in the vertical direction than horizontally. These deposits are usually very permeable and generally yield relatively large amounts of groundwater. In fact, outwash is predictably the most desirable source of municipal and domestic water in the Northeast. Commercially, this material is also valuable to the aggregate and concrete industries. In terms of hazardous waste management facility siting, this is usually one of the least desirable siting media, both in terms of groundwater protection considerations and the added expenses of construction of facility components to ensure isolation of the waste and containment of any leachate.

4. *Ice contact deposits.* These deposits are broadly the most varied in composition of all the glacially originated materials. Material ranging from clay to boulders may be present in one mass. However, ice contact deposits generally have not been consolidated and have fairly high permeabilities. When low in clay content, leachate attenuation

**Figure 6** Ablation till excavated from a field exploration test pit. This material lacks the silt and clay-sized matrix binder and lies at the angle of repose typical of a cohesionless material.

**Figure 7** Loose, stratified outwash sands of a kame delta; a cohesionless glacial unit recognizable by its broad and undulating landform.

capabilities are negligible. Ice contact deposits are usually not widespread and have the least distinctive geomorphic form.

5. *Alluvial deposits.* Alluvium is typically poorly graded, coarse, relatively permeable, and generally not particularly suitable for hazardous waste facility siting. Alluvium is often an attractive medium for large-scale groundwater development. Geomorphically it is not widespread and when present appears in valleys and other low-lying areas of higher average groundwater levels.

6. *Eolian deposits.* Eolian deposits of the Northeast are usually the fine sand and silt blankets of wind deposition along the margins of the Great Lakes, some proglacial lakes, or the margins of wider valleys. Although thin, they are widespread throughout much of New England. Strictly speaking, loess is also of eolian origin but is rarely found in sufficient thickness to be identifiable in broad-area searches or to be of much consequence to actual facility design or construction. In terms of facility siting, eolian deposits are not particularly favorable because of relatively high permeability and negligible clay mineral content.

7. *Beach deposits.* Beach deposits are found on a limited scale in the Northeast. These very fine to fine sands were deposited along glacial lake fronts and postglacial marine shorelines. These deposits are usually moderately dense and of moderate to low permeability. The areal extent of such sands is limited as is their thickness. The deposits should be identified during the broad-area search because of the possibility of site desirability when topographic conditions are favorable to hydrogeologic isolation, grain size is predominately fine, and permeability relatively low. Such sites may also be attractive for siting processing facilities which feature residuum waste disposal elsewhere.

8. *Glaciolacustrine deposits.* These materials (Fig. 8) are generally second only to lodgement till in terms of suitability for hazardous waste management facility siting. Deposited in areas once covered by glacial and postglacial lakes and ponds, these deposits are characterized by high percentages of clay and silt-sized particles. They are generally relatively low in permeability and high in leachate attenuation characteristics. However, exploration for these deposits must carefully search for the presence of sand-rich pockets

**Figure 8** Glaciolacustrine deposits vary considerably from location to location but, as shown here, are often rich in the clay and silt-sized fraction and can provide good containment characteristics when isolated from groundwater. Soils of this type, often 80 to 100% passing the No. 200 sieve, make excellent low-permeability bottom liner and final cover material candidates (metric scale).

and beds as variants in the depositional process. Sometimes, when referenced as parent material in U.S. Soil Conservation Service (SCS) county soil surveys, such deposits may also contain abutting or interfingering outwash sands, particularly in the instance of the smaller lacustrine bodies. Lacustrine deposits are found throughout the Northeast and are generally thick enough and of sufficiently broad areal extent to warrant consideration during broad-area searches. The general suitability for facility siting within such deposits is very high. Aside from the problem of detecting interbedded sands and silty sands, present topographic condition may be the single most influential facility siting factor.

9. *Organic deposits.* Organic deposits – materials in a state of decomposition in a silt and clay matrix – are generally found in areas of low topographic relief. Although generally cohesive, organic deposits are almost always unsuitable for facility siting, primarily on the bases of topographic setting and highly compressible foundation conditions. Being in low-lying areas, these deposits are generally saturated with groundwater or lie so close to the local groundwater table as to be highly susceptible to leachate contamination. In low-lying topography there is the added problem of encountering groundwater intrusion during periods of high precipitation.

## GEOMORPHOLOGY IN SITE QUALIFICATION

Environmental engineers use broad-area search overlays (such as Fig. 2c) in their decision-making process of the best trade-off between waste generation centers, short and safe haul routes, demographic considerations, environmental site suitability, and least cost aspects of construction at particular sites.

Any site chosen for future development into a waste management facility must be thoroughly investigated for its geologic, geotechnical, and hydrogeologic characteristics.

Among engineering geologists, an integrated system of alphabetic mapping symbols is gaining favor. This system, originally promulgated by Galster (1977), combines a parent symbol (such as G, denoting glacial origin) with a subscript defining the basic depositional process or environment (such as tl, for lodgement till) and a notation of engineering soil type in parentheses (such as scm, for sandy, clayey silt). The engineering soil symbol identifier is taken from the U.S. Army Corps of Engineers Unified Soil Classification System, a standard of practice among engineering geologists and geotechnical engineers. In the example cited above, the final map symbol becomes $G_{tl(scm)}$ and this portrays at once an expected range of important physical characteristics to the experienced engineering geologist. Galster's scheme is open-ended in that it may be modified to suit individual physiographic provinces, and this must usually be done.

Hazardous waste management facilities are effective only if they are designed and constructed so as to contain the objectionable components of the waste. For this reason, geologic studies that go into locating and designing the facilities must be conducted both thoroughly and carefully so as to define the physical properties of each geologic unit involved and to delimit the areal extent and general depth of each unit. At the same time, there is an upper economic bound on the amount of money that may be spent in securing this information. Since most sites are covered to some degree by surficial overburden, the full spectrum of geomorphic interpretation techniques should be used to extend available exploration resources at any given site as well as to help to define the most suitable of a number of sites. For purposes of hazardous waste management facility siting, geomorphic indicators of the origin of landforms can be used readily to identify the general physical property nature of the underlying engineering soil units. These units and property have been cited in Tables 1 to 3 and most geomorphologists will note how easily they can assist in these identifications based on their own regionally specific experience. The basic geomorphic interpretation as to landform type should be made as the first level of field mapping (Fig. 9) and before any subsurface explorations are undertaken.

Geologic contacts are then verified by backhoe pits and material properties further verified by exploratory borings. As a related cost-reducing measure, most of the boreholes are converted to groundwater observation wells, upon completion of drilling. Test pits and borings are also frequently converted to in situ permeability tests. Seismic refraction traverses are occasionally conducted between boreholes if the site design concept calls for considerable grading (as was the case in the area shown in Fig. 10) and bedrock is close enough to the ground surface to be of consequence. As an aside, surface rock excavation in the Northeast typically costs $8 to $27 per cubic meter at the present time.

In addition to the use of geomorphically derived symbols, the site often must be subdivided into geomorphic areas for the purpose of describing the relationships between geologic units and their physical properties, topography, groundwater, and bedrock. Figure 11 is an example of a site in central Massachusetts which has been geomorphically area subdivided for design-related purposes.

Geomorphic analysis of this site was performed with the goal of determining the presence and general paths of movement of groundwater and to find a suitably dense host material (lodgement till, in this case) which lay above the site piezometric level and which would be topographically isolated from potential future groundwater intrusion. Facility design features included provisions for sealing against leachate seepage, and site geology was being utilized to protect the facility further from contact with groundwater.

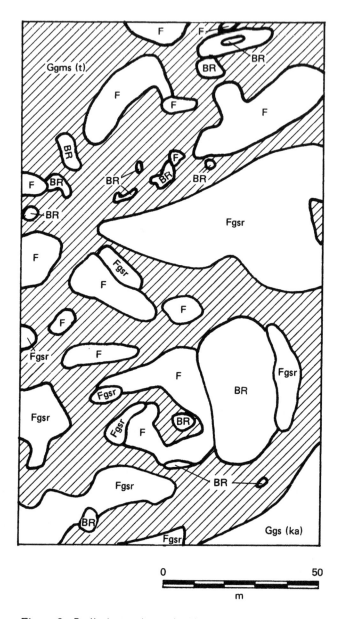

**Figure 9** Preliminary site review by geomorphic mapping, southeast Massachusetts. The engineering geologic map symbols are as follows: Ggms(t) is gravelly, silty sandy lodgement till; Ggs(ka) is gravelly sand of kame delta; F is undifferentiated fill (sand and gravel pit waste or spoil); Fgsr is fill made up of gravelly sand and rock; BR is bedrock.

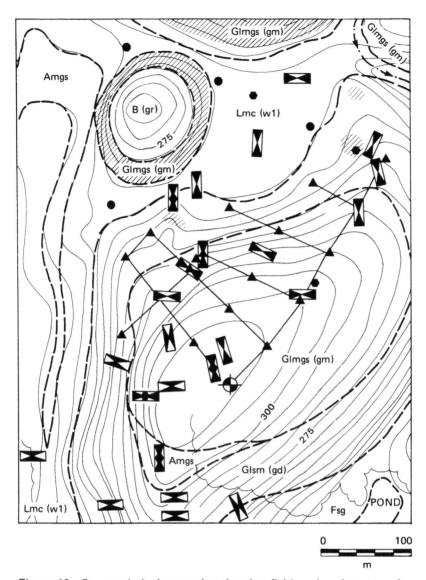

**Figure 10** Geomorphologic mapping showing field explorations at a site in southern Maine. The engineering geologic map symbols are as follows: Amgs is silty, gravelly sand alluvium; Lmc(wl) is water-lain (lacustrine) silty clay; Glmgs(gm) is silty, gravelly sandy ablation till (ground moraine); Glsm(gd) is sandy silt lodgement till (drumlin); B(gr) is granitic bedrock. Areas shown as hatched represent near-surface bedrock as estimated by the field geologist. Map symbols are as follows: rectangle with hourglass in-filling is an exploration backhoe pit; the hexagon denotes emplacement of a simple, slotted-casing, observation well; a solid circle represents a hand auger probe; the solid triangle is the location of machine-drilled (auger) borings along the solid lines representing seismic refraction traverses; and the quadrant-divided circle is a deep (13-m) auger boring with observation well installation.

69

## CONCLUSIONS

Hazardous waste management facility siting and design is among the most difficult of assignments facing the engineering geologist. Not only must the site geologic and hydro-geologic characteristics meet state and federal siting guidelines, but accurate assessment of the long-term interrelationships between the body of waste and the local groundwater regime must be made. Sites requiring extensive (expensive) engineered design features to mitigate poor site characteristics should be questioned as to their overall desirability for licensing. A thorough understanding of site and site-area geomorphic units and processes is essential in buiding a framework of sound geologic data to which geotechnical engineer-

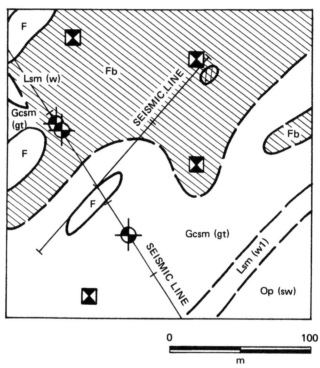

**Figure 11** Geomorphologic mapping applied to a site in central Massachusetts. The engineering geologic map symbols are as follows: F is undifferentiated borrow-pit spoil generated at the time the site (a drumlin) was stripped for use as fill in the adjacent interstate highway; Fb represents extremely bouldery borrow-pit spoil; Gcsm(gt) is clayey, sandy, silt lodgement till; Lsm(wl) is water-lain sandy silt; Op(sw) are organic (peat) deposits of a swamp, and; the small, circular, crosshatched outcrop is bedrock. Map symbols are as follows: squares with hourglass in-filling are exploration backhoe pits; quadrant-divided circles are hollow-stem auger and rock cored, piezometer-set and bentonite-backfilled observation wells; and the straight, solid lines are seismic refraction traverses.

ing properties and hydrogeologic condition assessments may be related, assessed, and design compensated.

## ACKNOWLEDGMENTS

Our thanks go to the Massachusetts Division of Water Pollution Control, the New York Environmental Facilities Corporation, the Town of York, Maine, and the Lewiston-Auburn (Maine) Water Pollution Control Authority for permission to reproduce examples of project-related mapping by the authors.

## REFERENCES

Galster, R.W. 1977. A system of engineering geology mapping symbols. *Assoc. Eng. Geologists Bull.* **14**(1), 39–47.

Roberts, K.J. and D.A. Sangrey 1977. *Attenuation of inorganic landfill leachate constituents in soils of New York.* School of Civil and Environmental Engineers, Cornell Univ., Ithaca, N.Y., Geotech. Rept. 77-2, 183 p.

U.S. Environmental Protection Agency 1980. Draft final hazardous waste and consolidated permit regulations. *Fed. Register* **45**(98), bk. 2, May 9, 33063–285.

# 5

# GEOMORPHIC MANIFESTATIONS OF SALT DOME STABILITY

*Z. Berger and J. Aghassy*

## ABSTRACT

The U.S. Department of Energy is presently seeking to identify salt domes that are suitable for high-level nuclear waste storage. In an effort to determine the tectonic and hydrologic stability of several salt domes in the Interior Gulf Coast, an investigation of their topographic expression was conducted. A model describing the erosional evolution of salt dome topography was devised and tested on 27 salt domes. As part of the model, it was postulated that erosional processes near salt dome areas are dominated by the subtle topographic highs that are formed over the salt diapirs. These topographic highs are attributed to differential loading and compaction of the underlying sediments. The model describes the assemblages of slopes and drainage elements as they progressively develop over and around tectonically and hydrologically stable salt domes. The model provides a conceptual framework for systematic analysis of salt dome topographies and thus aids in detecting salt domes that are suspected to have some tectonic and hydrologic instability problems.

## INTRODUCTION

Salt dome geology has been the subject of study for over a century (Ochsenius 1888). A wealth of data has been accumulating from drill and geophysical investigation in the Gulf of Mexico as well as in other parts of the world. The data served for present-day understanding and interpretation of the nature, age, origin, form, and size of domes (Atwater & Forman 1959; Trusheim 1960; Halbouty 1967; Mattox 1968; Braunstein & O'Brien 1968; Kehle 1971; Texas Bureau of Economic Geology 1980; and the five symposia on salt of the Northern Ohio Geological Society, 1960, 1963, 1970, 1974, 1980). This paper applies suitable geomorphic investigation means for the study of the surface expression of salt domes (or other structures), working through a hypothetical model which describes the evolution of salt dome-related topographies and may be useful in detecting tectonic or hydrologic instability problems.

The present paper was developed as a part of the U.S. Department of Energy (DOE) effort to identify salt domes which are suitable for nuclear waste storage in the Interior Gulf Coast. The report's main thrust is to present methods and standard procedures which were applied over a sample of domes and found applicable for the classification and evaluation of dome-related topographic evolution in soft sediment/low relief environments. The major elements utilized in this procedure rest on drainage components and slope categories as analyzed from a variety of maps and multi-altitudenal imagery. The paper suggests a uniform terminology and classification to be used for salt dome geomorphic analysis. Generic elements found in dome topographies are defined in the text and exemplified with the appropriate terminology in a series of figures and photos.

To establish a conceptual framework for the standard analysis of dome topographies, a model of the evolution of superdome topography was devised and tested on 27 domes in the Interior Gulf Coast. These domes create subtle relief in soft sediments requiring attentive observation of surface forms for recognition and identification. Therefore, a clear definition of basic slope and drainage elements as outlined by such a model can make superdome geomorphic analysis systematic and feasible.

## THE MODEL APPLIED TO DOME STUDY

Twenty-seven domes in the Northern Gulf Coast Salt Dome Province were examined through a variety of geomorphic observations and analyses. The landforms observed were categorized and the following model has been hypothesized to link the geomorphic evidence to structural elements related to salt domes.

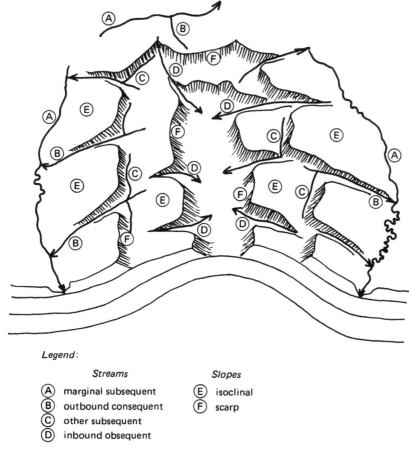

Legend:

|  | Streams |  | Slopes |
|---|---|---|---|
| Ⓐ | marginal subsequent | Ⓔ | isoclinal |
| Ⓑ | outbound consequent | Ⓕ | scarp |
| Ⓒ | other subsequent |  |  |
| Ⓓ | inbound obsequent |  |  |

**Figure 1**  Slope and drainage basic components over domal topography.

## Definitions

Analysis of the sample used for this study points to three basic stream and slope categories which are persistent throughout the different stages of dome-related topographic evolution. These need to be defined for uniform terminology as follows:

    (a) *Marginal subsequent streams.* These form as preexisting major streams and stream segments tend to adjust their flow around the newly formed round obstacle. They form a general circular pattern which outlines the outer perimeter of the dome (A in Fig. 1).

    (b) *Outbound consequent streams.* These form in a radiating pattern from the dome center. They may be either collected by the marginal subsequent components or flow directly outward (B in Fig. 1).

    (c) *inbound obsequent streams.* These streams flow toward the dome center (D in Fig. 1). They are usually collected by the central breaching stream following inversion of topography.

**Figure 2** Photo and interpretive sketch of southwestern Lampton dome, outlining isoclinal slopes and outbound consequent streams.

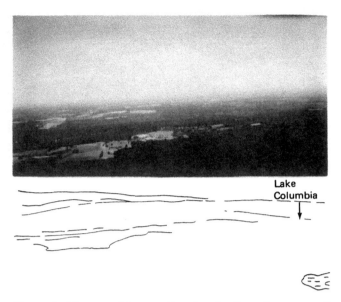

Lake
Columbia

**Figure 3**  Photo and interpretive sketch of northern part of
Lampton dome, outlining long isoclinal slopes.

(d) *Isoclinal slopes.* These are the long and gentle slopes or slope segments conform-
ing with the arched sediments (E in Fig. 1). These slopes may parallel the gently dipping
bedding planes (Fig. 2); however, they may also evolve into other inclined erosional
surfaces.

After these initial definitions a description of the model stages follows and further
definitions will be presented throughout the text at appropriate sections.

**Stage 1:**  *Positive relief stage.*  The surface configuration of the dome area exhibits
a central topographic high at this stage. From the dome center, long and gentle isoclinal
slopes extend down to the marginal subsequent streams with little or no interruptions
(Fig. 3). Radial drainage by outbound consequent streams is the most dominant element
over the major dome body and causes dissection of the long isoclinal slopes (Fig. 4).

The major marginal subsequent streams have adjusted to the newly formed environment; therefore, they exhibit gentle gradients and abundance of depositional features
along their courses (Fig. 4).

Lampton, McLaurin, and N. Richton domes in Mississippi as well as Central Butler,
Bullard, and Whitehouse domes in Texas exhibit most of the characteristics described
above and can be considered as domes of the positive relief stage. Figure 5a presents a
sketch map of Lampton dome outlining the different categories of its drainage components as verified by maps and photos.

Toward the end of this stage a new stream category begins to appear in the form of
tributaries to the outbound consequent streams, often meeting them at right angles but
mainly following the local strike orientation (C in Fig. 1). They are referred to as *other
subsequent streams* and are not numerous at this stage.

**Stage 2:**  *Breached stage.*  The increased erosional activity that occurs at the dome's
center results in a gradual lowering of the initial topographic high that existed there. This

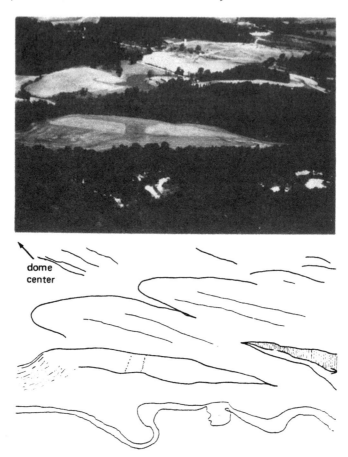

**Figure 4** Photo and interpretive sketch of southwestern Lampton dome, outlining deposition in marginal subsequent stream and dissection of isoclinal slope.

is done mainly by the headwaters of one or two dominant outbound streams which take over the drainage of the inner parts of the dome. An inversion of topography begins to shape up and a major depression develops at the dome's center (Fig. 6b). Slopes facing the central depression become numerous (F in Fig. 1). They will be referred to as *scarp-slopes*. These are shorter and steeper than the isoclinal slopes and face their opposite direction. Once these slopes exist, a new category of streams appears which follows the direction of the scarpslopes toward the central depression (D in Fig. 1). These are referred to as *inbound obsequent streams* and are destined to grow at the expense of the outbound consequent streams as they gradually capture increasing areas of their drainage basins.

The combination of scarpslopes carved out of the same stratigraphic unit may form a concentric rim around the dome central depression. Figure 7 portrays such rim as it appears in a three-dimensional display of digital terrain model (CALCOMP 1972). At times, differential lithological resistance of existing sedimentary units can result in several

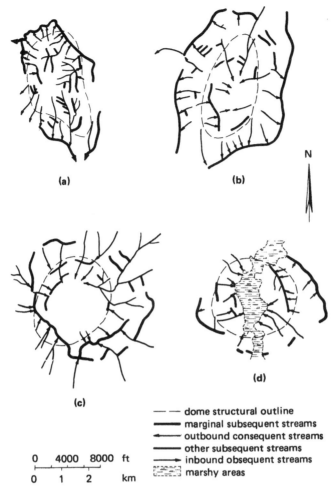

(a)

(b)

N

(c)

(d)

— — dome structural outline
▬▬▬ marginal subsequent streams
◄▬▬ outbound consequent streams
▬▬▬ other subsequent streams
▬▬► inbound obsequent streams
▭▭ marshy areas

```
0    4000   8000  ft
├────┼──────┤
0    1      2    km
```

**Figure 5** Drainage component sketches of selected domes.
(a) Lampton dome; (b) Keechi dome; (c) Palestine dome;
(d) Steen dome.

concentric rims, forming parallel ridges (Figs. 8 and 9). Subsequent streams separate between the different rims utilizing the weakness in contacts between different stratigraphic units.

A great number of domes exhibit the characteristics of the *breached stage*. Arcadia, Prothro, Chestnut, and Minden domes in Louisiana, Keechi and Mt. Sylvan domes in Texas, and Tatum dome in Mississippi exemplify the characteristics of this stage. A sketch of Keechi dome drainage components shows some of the characteristics of this stage (Fig. 5b).

**Stage 3:** *Obliterative stage.* As erosion continues the inbound obsequent streams expand headward beyond the rims and capture the better parts of the consequent outbound basins. The upper parts of the central breaching streams are no exception to this rule and in most cases end up dividing the dome into two halves as it widens its breaching central stream. This struggle results in substantial lowering of relief and rounding of forms

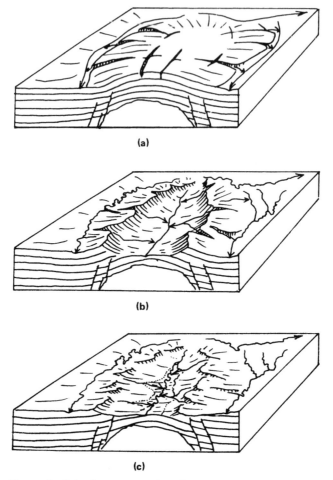

**Figure 6**  Salt dome normal surface evolution in soft sediment. (a) Positive relief stage; (b) breached stage; (c) obliterative stage.

leading to the obliteration of their original distinct outline (Fig. 6c). Extensive sediment yield of the inbound streams coupled with low gradient result in the widening of floodplain. At times, marshy areas form, especially when the outlet is restricted by rim relicts. At this stage a dominance of inbound obsequent streams rather than outbound consequent elements is quite outstanding (Fig. 5c), as exemplified by Palestine dome.

It is understood that in spite of their similar geologic history, different domal areas in the Interior Gulf Region are presently at different stages of their erosional evolution. Some of the possible reasons for that are the following:

(a) Depth of salt stock.
(b) Relative resistance of surface formation to weathering and erosional processes.
(c) Local and regional physiographic controls and relative position of dome to major relief elements.

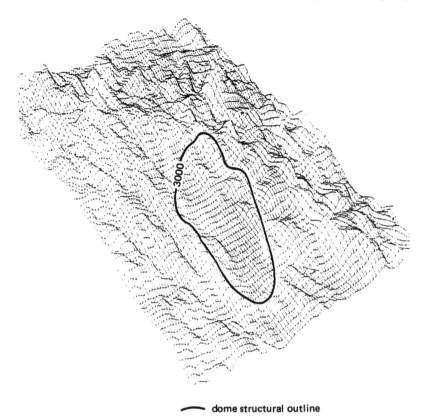

—— dome structural outline

**Figure 7** Three-dimensional display of digital terrain model of Cypress Creek dome (CALCOMP 1972).

Also, it is clear that domes can be in transitional states between the outlined stages in the model, thereby exhibiting a combination of characteristics of two stages.

Table 1 summarizes the occurrence frequency of geomorphic elements in salt dome topographies and can serve as a reference for a quick qualitative assessment.

Interruptions of the orderly sequence of erosional processes may occur at any stage of the model and are discussed in the following section. It is therefore not uncommon to find both the characteristics of a certain stage of the model and at the same time traits of certain type of disruption.

## COMPLICATIONS OF THE MODEL

During the testing phases of the model through analysis of numerous case studies, it became clear that geomorphic evolution of domes can be subjected to complications by the influence of either regional conditions or accidental processes. Some of the most predominant instances of such complications are related to physiographic regional influences. Other complications have been postulated and are related to possible dome stability. Details of both complication types are discussed below.

**Figure 8**    Photo and interpretive sketch of the northeastern half of Prothro dome, illustrating parallel ridges and concentric rims.

## Physiographic regional influences

The location of the dome in relation to the regional structure affects the erosional process and the resulting forms. In the particular case of the Interior Gulf Region, parallel cuesta scarps prevail with wide interfluve and bottomland in between. In cases where piercement occurs at main interfluve body or at bottomland location, surface manifestation appears to be normal and conforms with the characteristics described in the model. A departure from that is observed when the dome occurs at the cuesta scarp. Palestine dome in Texas exhibits some of the modifications due to such influences. It is important to note that the domal structures in turn also affect the cuesta. The results of these interactions can be summarized as follows:

(a) The retreat of cuesta scarp is increased in areas where domal structures exist. In most cases an embayment in the retreating scarp is formed around the dome (Fig. 10).

(b) The rims surrounding domes at cuesta scarp location are relatively conserved at the scarp side but highly obliterated at the opposite side. Figure 10 illustrates this phenomenon: rims at the eastern side of Palestine dome are quite evident, whereas only rim relicts remain on the western side. Divide migration through stream piracies between cuesta and dome obsequent elements highly accelerates rim obliteration.

(c) Obliteration at the opposite side of the scarp accelerates the elimination of the outbound consequent streams. This phenomenon is also evident in Palestine dome, where no outbound consequent elements exist on the western side (Fig. 5c).

**Figure 9** Photo and interpretive sketch of road crossing parallel structural steps at Richton dome.

It is true that such complication points to distinct departure from the proposed model. However, such departure should not be related to dome instability. Steen and Palestine domes in Texas, Kings dome in Louisiana, and Cypress Creek dome in Mississippi exhibit departures of this type due to their location in relation to cuesta scarps.

## Other complications

It has been postulated by scientists involved in the salt dome DOE project that two types of instability may occur: tectonic and hydrologic. Therefore, all candidate domes must be carefully examined for any sign of such disruptions by all possible means of inquiry, geomorphic tools included. The model in this case serves as a reference for normality and departures from its standard form not related to physiographic factors can point out to possible instability.

In the case of tectonic instability a rejuvenation of the drainage network is envisioned resulting in stream entrenchment and terrace evolution. Of the samples investigated in the present study, none seem to exhibit such a clearly detectable phenomenon.

As to the hydrologic-type disruption, which assumes dissolution of the salt and subsequent collapse, the presence of an extensive swampy area can be considered as one

Table 1    Occurrence frequency of stream and slope categories over Interior Gulf Region salt domes under stable condition and normal erosional development

| Category | Developmental stage | | |
|---|---|---|---|
| | Positive | Breached | Obliterative |
| Marginal subsequent | High | High | Medium |
| Outbound consequent | Very high | Medium | Very low |
| Inbound obsequent | Very low | Medium | High |
| Other subsequent | Low | Medium | Medium |
| Isoclinal slopes | High | Medium | Low |
| Scarp slopes | Low | High | Low |

surface manifestation of this type. Steen dome in Texas exhibits such attributes (Fig. 5d) and can be suspected of possible instability (Netherland, Sewell, and Associates, Inc. 1975).

## UTILIZATION OF SLCIR

In the present report, several standard geomorphic research techniques were used and yielded satisfactory results. However, it is important to mention the particular fruitfulness of the adaptation of low-altitude, side-looking infrared photography. This type of photography appears to be best suited in detecting surface expression of structural elements in soft sediment/low relief/heavily vegetated environment. Low-cost reconnaissance flight in a light aircraft, using a standard 35-mm camera and domestic color infrared film at 150 m above ground level provides significant clues for identification and differentiation between various slope types and channel elements inherent in several minor and major structures. This technique can successfully be utilized for in-depth investigation of particular domes as well as exploration for unknown ones. Other aspects in salt dome research, such as surface lithology, hydrology, and sedimentation, can benefit by the adoption of this technique.

## CONCLUSIONS

The results of the present study point toward the feasibility of applying geomorphic approaches for the investigation of salt domes. This is done by means of establishing the normal outcome of undisrupted erosional processes over domal topography. Once the standards are established, domes in question are tested, and those exhibiting departure from the standard are carefully scrutinized.

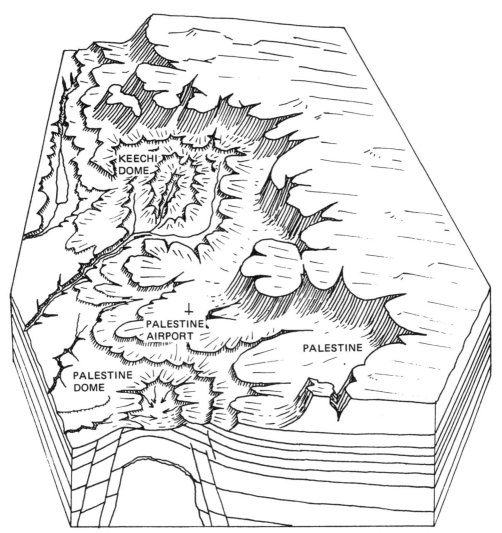

**Figure 10** Drainage and slopes assemblage resulting from the position of Keechi and Palestine domes in relation to the major cuesta.

This study followed these guidelines in testing Interior Gulf Region domes in Mississippi, Louisiana, and Texas. At this stage of the investigation most of the domes seem to exhibit either normal and orderly configuration regardless of stage, or bear some departures resulting mainly from physiographic cuesta-related complications rather than indications of instability.

It is the writers' opinion that the establishment of standard investigative procedures for describing Northern Gulf Coast Geomorphic expression is applicable in the exploration for salt domes as well as other geologic structures in areas of low relief, soft sediment, and heavy vegetative cover. The low-altitude SLCIR utilized here seems to be a highly promising and fruitful technique.

# REFERENCES

Atwater, G.I. and M.J. Forman 1959. *Nature and growth of southern Louisiana salt dome and its effect on petroleum accumulation.* Am. Assoc. Petroleum Geologists Bull. **43**, 2592–2622.

Braunstein, J. and G.C. O'Brien (ed.) 1968. *Diapirism and diapirs.* Paper presented at a symposium, New Orleans, La., 1965. Am. Assoc. Petroleum Geologists Mem. 8, 444.

CALCOMP 1972. *3D, a perspective drawing system.* Anaheim, Calif.: California Computer Products, 50 p.

Halbouty, M.T. 1967. *Salt domes; Gulf Region, United States and Mexico.* Houston: Gulf Publishing Company.

Kehle, R.O. 1971. *Origin of the Gulf of Mexico.* Unpub. ms., Univ. Texas at Austin, Geol. Lib., call no. q557 K260, unpaged.

Mattox, R.B. 1968. *Saline deposits.* Symposium, Geol. Soc. America Spec. Paper 88, 701 p.

Netherland, Sewell, and Associates, Inc. 1975. Preliminary study of the present and possible future oil and gas development of areas immediately surrounding the interior salt domes, upper Gulf Coast salt dome basins of east Texas, north Louisiana, and Mississippi, as of December 1975. Prepared for Union Carbide Corp., Nuclear Division, Oak Ridge National Laboratory, ORNL/SUB-75/87988.

Ochsenius, D. 1888. *On the formation of rock salt beds and mother liquor salts.* Proc. Acad. Natl. Sci. Philadelphia, 181–7.

Texas Bureau of Economic Geology 1980. *Annual report for 1979.* Bur. Econ. Geology, in preparation for Dept. of Energy Office of Waste Isolation.

Trusheim, F. 1960. *Mechanism of salt migration in northern Germany.* Am. Assoc. Petroleum Geologists Mem. 8, 1519–40.

# 6

# SLOPE MOVEMENTS RELATED TO EXPANSIVE SOILS ON THE BLACKLAND PRAIRIE, NORTH CENTRAL TEXAS

*James T. Kirkland and James C. Armstrong*

## ABSTRACT

When the subject of slope movement is discussed, geomorphologists rarely consider regions such as the Blackland prairie as problem areas. This is unfortunate because, owing to a combination of wet–dry climatic cycles and the presence of highly expansive montmorillonite clay soils, movements even on very gentle slopes are responsible for considerable damage and destruction to property.

Slope movements related to expansive soils are generally of two types: (a) those from soil creep and (b) rotational slides. Soil creep is due to the shrinking and swelling of the clay soils with seasonal moisture changes. Rotational slides are noticeable particularly along oversteepened road cuts, although they also occur on natural slopes. Two aspects of the clay soils help contribute to the formation of rotational slides: (a) the formation of large desiccation cracks provides easy access of surface waters to the soil, where they are adsorbed and help overload slope; and (b) the desiccation cracks often form what eventually becomes the head of the rotational slip surface.

## INTRODUCTION

Geomorphologists rarely consider regions such as the Blackland Prairie in Texas as geomorphic problem areas (Fig. 1). Even though the Blackland Prairie is well known for its expansive soil problems which are responsible for millions of dollars damage to homes, roads, and other structures annually, this is rarely thought of as a geomorphic problem. Nationwide, damage due to expansive soils exceeds $2·3 billion a year as reported by Jones and Holtz (1973). The present figure is probably double that and according to Mathewson et al. (1974) is twice the average annual loss from earthquake, hurricanes, tornadoes, and floods combined.

In regions such as the Blackland Prairie even very gentle slopes can undergo downslope movement due to soil creep and steeper slopes such as found along highways in earth embankments and along erosion channels are subject to rotational slumps. In fact, slumping may be one of the major mechanisms in the slow retreat eastward of the White Rock Escarpment that divides the Blackland Prairie into eastern and western subdivisions.

It is essential that a knowledge of geomorphology be combined with a practical background in geotechnics if these processes are to be understood. Indeed, geomorphol-

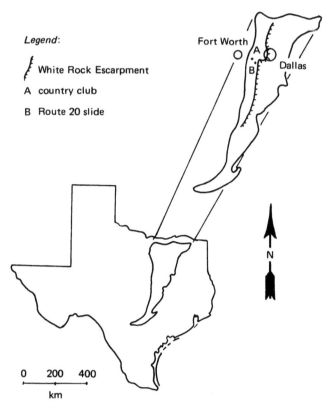

**Figure 1**   Location of the Blackland Prairie.

ogists should start looking more to civil engineering as a data base from which they can expand their understanding of process and in turn contribute to the solution of applied problems.


## PHYSIOGRAPHY

The Blackland Prairie has been subdivided by Winton and Adkins (1919) into two sections, the Eagleford subdivision (now written Eagle Ford) and the Austin chalk subdivision to the east. The two subdivisions are separated by the White Rock Escarpment capped by the Austin chalk. The escarpment trends roughly north–south along the west side of Dallas. The strata are all of Cretaceous age and dip about 2° to the southeast toward the Gulf Coast. Except for the White Rock Escarpment the region consists of a gently rolling treeless surface. This paper is primarily concerned with the Eagle Ford subdivision of the Blackland Prairie.

According to Gould (1962), the region is classified as a "true Prairie" and has little bluestem as a climax dominant plant, although most of the region is now either urbanized or under cultivation. Very little of the Blackland Prairie retains the climax vegetation. Mesquite has invaded the prairie in recent times, and various domestic trees such as mulberry, elm, and pecan have been planted in urban areas.

Table 1  Clay mineral percentages determined by x-ray diffraction on Eagle Ford soil, Grande Prairie, Texas (after Poor 1976)

| Depth (ft) | Illite | Montmorillonite | Kaolinite | Chlorite |
|---|---|---|---|---|
| 0·0–1·0 | 34·6 | 27·9 | 29·5 | 8·0 |
| 1·0–2·0 | 29·4 | 32·7 | 30·0 | 7·9 |
| 2·0–3·0 | 32·8 | 21·9 | 35·8 | 9·5 |
| 3·0–4·0 | 25·0 | 37·5 | 37·5 | 0·0 |
| 4·0–5·0 | 37·2 | 25·3 | 37·5 | 0·0 |
| 5·0–6·0 | 34·7 | 26·4 | 38·9 | 0·0 |
| 6·0–7·0 | 33·3 | 25·7 | 33·2 | 7·8 |
| 7·0–8·0 | 50·6 | 11·0 | 31·6 | 6·8 |
| 8·0–9·0 | 28·0 | 35·0 | 37·0 | 0·0 |
| 9·0–10·0 | 32·2 | 28·0 | 30·2 | 9·6 |
| 10·0–12·0 | 34·6 | 14·1 | 38·0 | 11·5 |

## SOILS

Soils on the Eagle Ford subdivision of the Blackland Prairie between Dallas and Fort Worth are mostly Vertisols belonging to the Houston Black Association (Soil Conservation Service 1972). These soils are well known among soil scientists and engineers for their "severe" shrink-swell potential, very low permeability, and severe corrosivity characteristics. Houston Black soils are residual soils derived from weathering of the Eagle Ford shale and have natural slopes ranging from 0 to 8°. When wet, the soils are stiff and "waxy" but crumble into very small blocky units upon drying. Wet soils are so plastic and sticky that the natives of the region often refer to themselves as "growing taller" when walking across fields after a rain.

Houston Black soils are high expansive; that is, they display a large capacity for volume change with relatively small changes in moisture content. These volume changes most often occur within the upper 5 to 10 m or within the "active zone." These swelling soils are capable of exerting tremendous expansion pressures on structures in contact with them. Krynine and Judd (1957) use the expansion pressures from Eagle Ford soils as an example of the extreme pressures that occur in expansive soils and give a value of 14·65 kg/cm$^2$ for such pressures. Even this figure is low, as will be noted in the section on engineering characteristics. These expansion pressures are due to the relatively large montmorillonite content in the soil combined with the seasonal wet–dry climate cycles characteristic of the North Central Texas region. Clay mineral percentages for Eagle Ford soils are given in Table 1.

## ENGINEERING CHARACTERISTICS

The surficial clay soils weathered from the Eagle Ford shale possess liquid limits ranging from percentages in the upper 40s to the 80s, plastic limits ranging from percentages in the upper teens to the upper 30s, and plastic indexes ranging from the lower 30s to the mid-60s. Swell tests performed on soils from the Dallas–Fort Worth metroplex have registered swell pressures in excess of 30·27 kg/cm$^2$ and free swell percentages in excess of 15%. Vertical movement in these soils depends to a large degree upon the location of the permanent groundwater table. Vertical movement in excess of 10 cm is quite common, with occasional movements of 25 to 30 cm. Severe damage resulting from excessive heave and/or shrinkage is commonplace in structures found on or in the active zone of the Eagle Ford clay soils.

The Eagle Ford clay soils are highly fractured, often showing desiccation cracks in excess of 7 cm wide and extending to varying depths below the surface. The Eagle Ford soils also contain slickensides throughout the soil profile attesting to the movement of these soils during shrink–swell cycles (Fig. 2). These characteristics have been observed in numerous subsurface investigations throughout the Blackland Prairie as well as other areas with highly expansive soils.

Loss of shear strength with time, in slopes formed from and in overconsolidated soils such as the Eagle Ford clay soils is well documented (Skempton 1964; Bjerrum 1967). As a slope is cut into natural soil, there is a loss of confining pressure within the soil mass. The loss of confinement in conjunction with moisture allows expansion and a build-up of negative pore water pressure along the potential failure zone. To satisfy the negative pore water pressure conditions, water tends to migrate toward the zone of low pressure, thus saturating the soil and decreasing the shear strength. When the shear strength along the potential failure zone reaches a magnitude equal to that of the shearing stress applied by gravity forces within the slope failure results. The final shear strength resulting from the change in pore water conditions is much smaller than that usually reported in geo-technical reports that are based upon unconfined compression test data. The soil in the slope fails at a stress approaching the residual strength as shown in Figure 3. This is the reason that many slopes in residual soils fail years after they are formed. The change in strength from peak to residual can result in over a 50% loss in strength with time.

## SLOPE MOVEMENT

### Soil creep

Soil creep has been defined by Sharpe (1938) as any movement that is imperceptible, except by measurements over long periods of time. Soil creep is the response to a down-

**Figure 2**  Slickensides in Eagle Ford shale. Test pit was located at the Dallas–Fort Worth Airport.

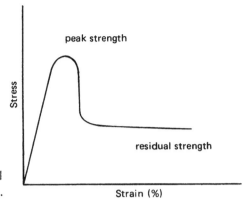

**Figure 3** Shear characteristics of a typical overconsolidated clay (after Bjerrum 1967).

hill shear stress and has been classified as continuous creep and as seasonal creep by Terzaghi (1950). Continuous creep is the response of materials with rheological properties of clays to a constant downhill stress. The behavior of materials undergoing continuous creep are well described by Carson and Kirkby (1972) and Kojan (1969).

Seasonal creep, on the other hand, is related to seasonal variations in soil micro-climate. The classic model for seasonal creep involves the freezing and thawing of water, and assumes that expansion due to freezing is perpendicular to the slope and contraction is in a more vertical direction, thus producing a zigzag downhill movement. This is effective only if soil strength is large during expansion and slight during contraction in order to produce a nonrecoverable downslope displacement. In actuality, movement is probably more U-shaped than zigzag (Carson & Kirkby 1972; Kojan 1969). Nonetheless, a satisfactory formulation of the basic creep mechanism is still lacking according to Kojan (1969). This is particularly true in the case of overconsolidated residual soils such as those formed on the Eagle Ford shale. In his study of the South Bosque shale, part of the Eagle Ford Group in the Waco area, Font (1976) estimated that the heavily overconsolidated nature of the shales reflected the removal of 150 to 600 m of overburden. This results in a condition whereby slope failure can occur under lower shear stresses than indicated by peak laboratory shear strength tests.

It should be reemphasized that shrink–swell cycles by themselves are inadequate to produce soil creep because vertical movements downhill only retrace those made during the upward part of the cycle. Any increments of strain must be produced by shear stresses parallel to the slope, and the slope and these shear stresses must be great enough to induce a certain amount of shear failure. Soils formed on Eagle Ford shales have very high liquid limits and thus can be expected to be highly expansive. As the soils increase in moisture content and approach the liquid limit, they pass from a plastic state and approach a liquid state, thus decreasing in shear strength, and become more susceptible to soil creep. Additional problems are created by the load exerted by structures built on such slopes, thereby increasing the effective shear stress during shrink–swell cycles. The effects of this type of creep are readily apparent in the form of cracked and separated foundations and tilted retaining walls.

It is likely that slopes, particularly under a load, also undergo a certain amount of plastic flow in addition to those movements that are the result of shrink–swell cycles. Certainly the most noticeable effects of movement of structures are easily correlated with seasonal variations in soil moisture content.

A special case of soil creep involves the added effect of transpiration by trees. Blackwell et al. (1979) found that a 15-m cottonwood tree could produce 2·3 cm of surface withdrawal in 1·75 days and 8·3 cm withdrawal in 28 days. Soil desiccation by transpiration is often so extreme at foundation corners as to produce enough slope in itself to induce foundation failure and separation. This process is particularly devastating when repeated over several shrink–swell cycles. The effects of this type of accelerated soil desiccation is readily apparent when one examines the vast amount of damage to single-family dwelling structures located on expansive soils (Tucker & Poor 1978).

## Rotational slides

Rapid slope failure in the region is usually of the rotational type with well-defined cylindrical slip surfaces. These failures most often occur along highways and other areas where slopes have been artificially oversteepened (Fig. 4). Font (1977) distinguishes these from slumps where the shale is loaded by overlying Austin chalk. In the latter case deformation may be in "the form of slow earthflow" or where more rapid failure takes place cylindrical slip surfaces develop. Failures most often occur during intense rainfall when water penetrates desiccation cracks in the clay soil forming a perched water table (Salas 1965). This introduction of water not only increases pore water pressure, but also places an additional load on the slope. These desiccation cracks can also form the beginning of the failure plane.

Loading upon the potential failure plane is not uniform; therefore, part of the failure zone is stressed more highly than other areas. The mode of failure that results is termed *progressive failure*. This means that failure does not develop at all points along the potential failure surface at the same instant in time. Progressive failure occurs when an overconsolidated clay soil in a slope is strained to a point beyond which the peak strength can no longer exist (see Fig. 3). This results in a corresponding reduction in shear strength, and a buildup of stresses, in excess of the peak strength in another part of the potential failure zone. As time passes, this point too will be strained past its peak shear strength and overstressed. This phenomenon continues until a slip surface develops that possesses only residual strength. If the slope has been designed using strength values in excess of the residual, ultimately failure will result.

Figure 4 Typical rotational slump along the Dallas–Fort Worth Turnpike.

## EXAMPLES AND CASE HISTORIES

### Park Valley Country Club Pool, Grand Prairie, Texas

The Park Valley Country Club provides an excellent example of several geomorphic principles as well as two unusual examples of soil creep. The exact date of construction of the club pool is unknown, but it was constructed on a combination of Eagle Ford clay soil and fill. The fill is almost indistinguishable from the soil formed in situ except that it contains a few pebbles, probably introduced during construction. The swimming pool has undergone a downslope movement such that the south end of the pool is 50 cm below the north end (Fig. 5). Interestingly enough, the whole pool has moved as a unit without showing any signs of cracking. The question posed by the manager of the country club was: If they spent $12 000 to $14 000 on pool repairs to make it swimmable again, could the pool be expected to last another 4 to 5 years before it again became unusable due to further movement? The more important question that was not asked is: What was the cause of the movement and could anything be done to prevent further movement and yet not cause buckling of the pool?

Examination of the creek behind the pool revealed exposed tree roots, indicating that the creek had undergone fairly recent incision, creating excessively steep banks. Further study led to the conclusion that the entrenched stream channel was the result of channel straightening such that incision was the natural response of the stream channel as it adjusted to the new conditions imposed upon it (see Chapter 15).

Thus it was fluvial erosion that created the slope instability resulting in soil creep. The recommended solution was the construction of a retaining structure at the toe of the slope, thus effecting a permanent solution.

Interestingly, the pool was not the only object undergoing downslope creep movement. Examination of the adjacent tennis courts, constructed on 60 to 90 cm pad of fill, showed evidence of creep movement. Greatest movement was at the corners of the courts, where shrink–swell movements are greatest due to greater seasonal moisture changes. Failure of the concrete was in the form of a series of en échelon fractures (Fig. 6).

### Interstate Route 20 Slide

This slide is located along Interstate Route 20 just west of Dallas along the edge of the White Rock Escarpment. It is of particular interest not only because of its size, but also because it is what Font (1977) termed a stratigraphically controlled slope failure. Font describes two modes of failure for slopes where the shales are overlain by a chalk caprock: (a) slow earth flow of the caprock-loaded shales and (b) more rapid rotational

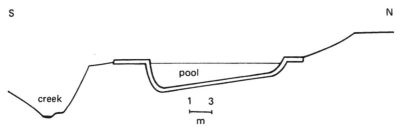

**Figure 5** Cross-sectional view of the Park Valley Country Club pool.

**Figure 6** En échelon fractures in the Park Valley Country Club tennis court. These were formed by soil creep associated with seasonal shrink–swell movements of the soil.

failures with well-defined cylindrical slip surfaces. This latter type of failure takes place following considerable strain. In the case of the Interstate 20 slide (Fig. 7) the design slope was 1:1. This created an excessively steep slope with a caprock thickness of 3 m of Austin chalk loading the slope. Failure occurred in the spring of 1977 releasing about 125 000 m³ of material that was removed by the highway department at a cost of about $330 000. The present slope has been reduced to 3·5:1 with two 3-m berms to control drainage.

Even though the slope failure was induced by construction, fossil evidence movement is present along the original slope of the escarpment less than 60 m away. Failure to notice this evidence resulted in the creation of an unsafe condition that only due to good fortune did not cause any traffic accidents. Failures in natural slopes similar to the Interstate 20 failure have been described by both Font (1977) and Allen (1977) and represent one of the natural modes of eastward escarpment retreat.

## CONCLUSIONS

Both the shrink–swell cycles and resulting desiccation cracks in expansive clay soils can be responsible for slope movements. The shrink–swell cycles of the clay soils cause a form of seasonal creep which can result in considerable damage to structures founded upon these soils. The desiccation cracks resulting from these shrink–swell cycles provide easy access of rainwater to the clay soils aiding in rapid moisture changes as well as forming planes of weakness along which rapid failure can occur.

Slopes in consolidated clay soils should be designed using the residual strength values rather than peak strength.

**Figure 7** The Interstate Route 20 slide. The picture was taken during the removal of slide material.

# REFERENCES

Allen, P. 1977. *Analysis of an escarpment hillslope for lane use planning in Mountain Creek Basin, Dallas-Ellis counties, Texas.* Dallas, 391 p.

Bjerrum, L. 1967. Progressive failure in slopes of over-consolidated plastic clay and clay shales. *Jour. Soil Mechanics Found. Div., Am. Soc. Civil Engineers* **93**, Pt. I, p. 3-47.

Blackwell, C.C., J.T. Kirkland, A.L. Blackwell, and L.L. Kirkland 1979. A mathematical model for predicting the dynamics of expansive soil desiccation due to transpiration of trees. Abs. Am. Soc. Adv. Sci. Ann. Mtg. Program with Abs., Houston.

Carson, M.A. and M.J. Kirkby 1972. *Hillslope form and process.* Cambridge: Cambridge University Press, 475 p.

Font, R.G. 1976. Relationship between the geologic history and engineering properties of two Cretaceous shales. *Texas Jour. Sci.* **27**(2), 267-84.

Font, R.G. 1977. Engineering geology of the slope in stability of two over-consolidated North-Central Texas shales. In *Landslides reviews in engineering geology*, D.R. Coates (ed.), Vol. 3, 205-12. Boulder, Colo.: Geological Society of America.

Gould, F.W. 1962. *Texas plants-a checklist and ecological summary.* College Station, Tex.: Texas A & M University Press.

Jones, E.C. and W.G. Holtz 1973. Expansive soils - the hidden disaster. *Civil Engineering* **8**, 49-51.

Kojan, E. 1969. Mechanics and rates of natural soil creep. *Proc. 5th Ann. Eng. Geology and Soils Eng. Symp.*, Idaho Dept. of Highways, Pocatello, Idaho, 233-53.

Krynine, D.P. and W.R. Judd 1957. *Principles of engineering geology and geotechnics.* New York: McGraw-Hill, 730 p.

Mathewson, C.C., J.P. Castleberry, and R.L. Lytton 1974. Analysis and modeling of the performance of home foundations on expansive soils in central Texas. *Assoc. Eng. Geologists Bull.* **12**(4), 275-302.

Poor, A.R. 1976. *Interim report of experimental foundation designs on expansive clay soils.* Construction Research Center, Univ. Texas at Arlington, Arlington, Tex.

Salas, J.A.J. 1965. *Calculation methods of the stresses produced by swelling clays, engineering effects of moisture change in soils.* College Station, Tex.: Texas A & M University Press, 330-40.

Sharpe, C.F.S. 1938. *Landslides and related phenomena.* New York: Columbia University Press, 136 p.

Skempton, A.W. 1964. Long term stability of clay slopes. *Geotechnique* **14**(2), 77-101.

Soil Conservation Service 1972. *General soil map, Tarrant County, Texas.* U.S. Department of Agriculture, Soil Conservation Service, Temple, Tex.

Terzaghi, K. 1950. *Mechanics of landslides in application of geology to engineering practice.* Berkey volume. Boulder, Colo.: Geological Society of America, 83-124.

Tucker, R.L. and A.R. Poor 1978. Field of study of moisture effects on slab movements. *Jour. Geotech. Eng. Div. Am. Soc. Civil Engineers* **104**(GT4), 403-14.

Winton, W.M. and W.S. Adkins 1919. *The geology of Tarrant County.* Univ. Texas Bull. 1931, 122 p.

# 7

# PREDICTION OF ENGINEERING PROPERTIES AND CONSTRUCTION CONDITIONS FROM GEOMORPHIC MAPPING IN REGIONAL SITING STUDIES

## Gary E. Christenson, James R. Miller, and Denise D. Pieratti

### ABSTRACT

Siting studies for the MX advanced ICBM program require a rapid and inexpensive method of evaluating construction conditions over large areas. The engineering evaluations important to MX siting are for road construction, excavation, and aggregate resources. Generalized evaluations of these parameters are greatly facilitated by geomorphic studies of terrain conditions, soil types, and induration of basin-fill deposits.

In the alluvial valleys of the Great Basin considered for MX siting, fluvial, eolian, lacustrine, and alluvial fan deposits have been differentiated according to grain size, relative age, and/or morphology. Map units thus directly address terrain conditions, soil type, and induration. Eolian and modern lacustrine (playa) deposits denote areas of predictable soil type and terrain. Fluvial, pluvial lacustrine, and alluvial fan deposits are highly variable in terms of soil type, induration, and terrain.

Sandy and gravelly alluvial fan and lacustrine deposits are most suitable for road construction from all standpoints except terrain, whereas eolian sands and fine-grained deposits are least suitable. Fine-grained lacustrine and alluvial fan deposits present fewest excavation difficulties, and cemented gravelly alluvial fan and lacustrine deposits present greatest difficulties. Sandy and gravelly lacustrine shoreline and alluvial fan deposits are most suitable for use in concrete aggregate but still require processing. Eolian sand and all fine-grained deposits are unsuitable as aggregate sources.

## INTRODUCTION

This paper presents a methodology for the use of geomorphology to predict engineering properties of basin-fill deposits within valleys of the Basin and Range Physiographic Province. The methodology combines standard techniques for evaluating engineering properties of geologic materials with geomorphic identification and mapping of basin-fill units. The methodology has been developed over a period of several years as part of geotechnical siting studies performed for the MX advanced ICBM program. It has been applied at many test sites throughout the Basin and Range Province and has proved to be a suitable systematic approach for predicting and characterizing engineering properties.

The primary purpose of this paper is to present the engineering properties of landforms and basin-fill deposits that have been determined through the mapping and sampling methodology. Geologic and geomorphic processes that influence the engineering properties of the basin fill will be discussed only to the extent needed to characterize

94

the properties. The application of these techniques to construction considerations is exemplified by one valley under consideration for MX siting.

## Siting parameters for MX

MX is to be a movable advanced ICBM system in which the missile and its transporter will move periodically among fixed shelters. These partially buried shelters are spaced several thousands of feet apart and number more than 4000. Roads connecting these shelters will be unsurfaced.

The broad alluvial valleys of the Great Basin are under consideration for siting of the MX system. These valleys are characterized by great thicknesses of unconsolidated deposits, predominantly alluvial fan and lacustrine sediments. In each valley, assessments of construction conditions for MX include:

(a) Mapping engineering parameters of basin-fill deposits for road design and assessment of excavation conditions.
(b) Identifying aggregate resources for concrete and road construction.

These assessments are made in the portions of the valley that are suitable for siting. Site suitability of a valley is defined by absence of rock, well-indurated basin fill or groundwater within 15 m of the surface to facilitate construction of shelters. Steep or highly dissected terrain that would inhibit movement of the MX vehicle must also be avoided.

## GEOLOGIC MAPPING

Geologic mapping is used to determine the extent of units present in a valley and provide the basis to extrapolate engineering properties and determine construction conditions. A unique mapping nomenclature was developed for the purpose of providing to the design engineer and planner, as well as the geologist, easily understandable data. All basin-fill map units are identified with the letter "A" to differentiate them from rock units. Basin-fill units are then further differentiated according to depositional environment. Depositional environments identified and their associated map symbols are fluvial (A1, A2), eolian (A3), lacustrine (A4), and alluvial fan (A5). These units are further subdivided according to grain size, relative age, and/or morphology. All units are differentiated according to grain size except for eolian deposits, which are homogeneously sand-sized. Grain size units are shown in Table 1 and conform to the Unified Soil Classification System (USCS). In addition to grain size distinctions, relative age differentiations are made

**Table 1**   Grain size unit descriptions

| Symbol | Predominant grain size | USCS symbols |
|--------|------------------------|--------------|
| f | > 50% fines (< 0·074 mm, passing No. 200 sieve) | ML, CL, MH, CH |
| s | > 50% sand (4·76–0·074 mm, passing No. 4 sieve, retained on No. 200 sieve) | SM, SC, SP, SW |
| g | > 50% gravel (> 4·76 mm, retained on No. 4 sieve) | GM, GC, GP, GW |

for fluvial deposits (modern channels and terraces), lacustrine deposits (modern playa and older lake deposits), and alluvial fan deposits (younger, intermediate, and older fans). Only morphology is differentiated in eolian sand deposits, which occur either in dunes or in uniform sheet deposits.

Similarly, composite map symbols were developed for rock units. All map symbols provide users of the map and other data with information on the physical characteristics of rock and basin-fill deposits that would not otherwise be presented if the more conventional map symbol nomenclature focusing on geologic age and formation characteristics had been applied. Table 2 illustrates the geologic unit symbols and summarizes the characteristics of basin-fill units based on a compilation of data taken at over 1250 locations in twenty valleys throughout the Basin and Range Province. Figure 1 illustrates the appearance of these units on aerial photography in Cave Valley, Nevada. Details of Table 2 and Figure 1 are discussed in the following sections.

## Fluvial deposits

Stream channel and floodplain deposits of significant areal extent are not common, owing to the predominantly closed drainage system. However, major streams occur locally connecting basins, flowing along basin axes between alluvial piedmonts, or flowing along major alluvial fans. Many of the major modern streams occur in the bottoms of wide, deeply incised Pleistocene drainageways locally greater than 30 m deep. Streams are generally incised into alluvial fan or older fluvial (terrace) deposits. Stream terraces commonly parallel modern channels at various heights.

**Figure 1**   Oblique aerial view to the north of Cave Valley, Nevada, illustrating the appearance of selected geologic units on aerial photography.

**Table 2** Derivation of geologic unit symbols and generalized characteristics of units

| Depositional environment[a] | Relative age or morphology[a] | Predominant grain size[a] (USCS) | Composite symbol | Grain shape[b] | B-horizon development | Stage of caliche development[c] | Average depth of incised drainages (m) | Average gradient/surface slope (%) |
|---|---|---|---|---|---|---|---|---|
| Fluvial (A1, A2) | Modern channel (A1) | Fines (f) | A1f | – | None | None | – | 0·4 |
| | | Sand (s) | A1s | Sub-ang | None | None | – | 0·7 |
| | Terrace (A2) | Sand (s) | A2s | Sub-rnd | None | I | 1·0 | 1·3 |
| Eolian (A3) | Sheet (s) | Sand | A3s | Sub-rnd | None | None | – | 1·4 |
| | Dune (d) | Sand | A3d | Sub-rnd | None | None | – | Variable |
| Lacustrine (A4) | Modern playa | Fines (f) | A4f | – | None | None | 0 | 0 |
| | Older (o) | Fines (f) | A4of | – | None | None | 1·2 | 0·9 |
| | | Sand (s) | A4os | Sub-rnd | None | None | 1·3 | Variable |
| | | Gravel (g) | A4og | Sub-rnd | None | I | 5·6 | Variable |
| Alluvial fan (A5) | Younger (y) | Fines (f) | A5yf | – | None | None | 0·5 | 0·6 |
| | | Sand (s) | A5ys | Sub-ang | None | I | 0·5 | 2·2 |
| | | Gravel (g) | A5yg | Sub-ang | None | I | 0·8 | 4·1 |
| | Intermediate (i) | Sand (s) | A5is | Sub-ang | Poor | II | 2·3 | 3·1 |
| | | Gravel (g) | A5ig | Sub-ang | None | II | 3·0 | 3·7 |
| | Older (o) | Sand (s) | A5os | Sub-ang | Poor | III | 7·9 | Variable |
| | | Gravel (g) | A5og | Sub-ang | None-poor | III | 4·2 | Variable |

[a] Symbols in parentheses combined to form composite geologic unit symbol.

[b] Grain shape: sub-ang, subangular; sub-rnd, subrounded.

[c] Stages in morphogenetic sequences of caliche development (Gile 1966): I, thin, discontinuous pebble coatings, filaments; II, continuous pebble coatings, some interpebble fillings; III, many interpebble fillings, nodules, internodular fillings; IV, laminar horizon overlying plugged horizon.

Terrain on active fluvial deposits is generally flat with local bar and channel topography. Terrain is also very flat in stream terraces, except along dissected scarps paralleling the modern stream course. Soils in stream channel and terrace deposits are very similar to those in alluvial fan deposits, but locally exhibit more rounding of grains. In many cases, the only means for distinguishing terrace deposits from alluvial fan deposits bordering an incised modern channel is by their down-valley gradient parallel to that of the modern stream. Mappable fluvial deposits are generally restricted to valley bottom areas where surface gradients are low and materials have already undergone considerable reworking and sorting in the alluvial fan environment. Thus fluvial deposits are generally more uniform in terms of grain size and are better sorted than alluvial fan deposits. Indurated caliche occurs locally on higher, older stream terraces, but is generally absent in all fluvial deposits.

## Eolian deposits

Eolian sand deposits commonly occur downwind (north and east) of playas either in dune fields or in uniform sheetlike deposits mantling the surface. Most eolian deposits are stabilized by vegetation under present conditions. Whether active or stabilized, eolian sand deposits represent uniform areas of predictable terrain and soil conditions. Terrain varies from essentially flat (sheet sands) to hummocky (dune sands). Soils are generally nonindurated, well-sorted, fine- to medium-grained sands of loose consistency. Eolian sands are highly permeable and have generally not been stabilized long enough to exhibit any soil profile development or caliche cementation.

## Lacustrine deposits

Lacustrine deposits include modern playas and older pluvial lake deposits. In most valleys, modern playas are circumscribed by concentric shorelines at higher elevations which represent strand lines of the Pleistocene lakes which occupied these basins. Modern playa deposits are uniformly flat and undissected, and deposits are fine grained (silts and clays) and uncemented. Older lake deposits, however, are highly variable in terms of both terrain and soil type. These deposits consist basically of shoreline deposits (A4og, A4os), lake basin deposits (A4os, A4of), or a combination of both within the zone of lake fluctuation (Fig. 1). Shoreline features range from constructional bars, spits, and beaches forming ridges and knobs to wave-cut notches cut in alluvial fan deposits. Fine-grained lake basin deposits are generally flat, but other lacustrine deposits are of variable slope, exhibiting various degrees of dissection.

The nature of shoreline deposits varies with the size of the lake, the types of materials delivered to the lake by streams, and the orientation of the shoreline with respect to wind direction and resultant wave action. Extensively reworked shoreline deposits of well-rounded and well-sorted gravel occur in valleys occupied by Lake Bonneville. Shoreline deposits in other valleys occupied by isolated, smaller Pleistocene lakes such as shown in Figure 1 are generally sandy and are not as extensively reworked. Lake basin sediments are fine grained (silts and clays). Deposits between shorelines and lake basin sediments are generally sandy and exhibit a component of alluvial reworking. Lacustrine deposits are generally nonindurated, although indurated tufa deposits occur locally in gravelly shoreline bars.

## Alluvial fan deposits

Alluvial fan deposits occur at the surface in over 75% of the area. They are very diverse and complex in terms of terrain, soil type, and induration. Most alluvial fans in the Basin and Range Province are composite fans that consist of deposits of various ages which represent responses to various climatic, tectonic, or other base-level changes through time. Alluvial fans consist of poorly sorted, angular, coarse debris-flow and water-laid deposits near the mountain front, grading into finer, predominantly water-laid deposits downfan (Bull 1978). Sorting and rounding of grains increases downfan. The rate of change in alluvial fan deposits downfan is a function of many variables (Bull 1978) and will differ in adjacent fans as well as in deposits of various ages of a single fan. Because of this, differentiation of grain size units in alluvial fan deposits is more difficult than in other types of deposits. To delineate grain size variations, field data are taken on the fan at various distances radially from the fan apex. The general zone where the predominant grain size changes from silt and clay to sand or from sand to gravel are determined. To map the extent of the deposits of each grain size, contacts are extrapolated across the fan at an approximately equal distance from the apex and at an approximate constant slope.

Grain size contacts also commonly follow contacts between adjacent fans. For example, gravelly deposits may occur much farther downfan on a fan with a large, steep drainage basin than on an adjacent fan with a smaller basin of low relief. Additionally, grain size contacts may follow contacts differentiating various ages of deposits on a single fan. When a fan surface becomes entrenched and the locus of deposition shifts downfan, material deposited on the resulting younger fan may differ from that in adjacent areas of the entrenched fan surface. For example, gravel may occur farther downfan on a relatively older fan than on a younger fan from the same drainage basin due to a decrease in source drainage basin relief with time or due to the effects of climate change (Bull 1978).

In this study, three relative ages of alluvial fan deposits have been differentiated. Table 3 presents photogeologic criteria used in differentiating various ages of alluvial fans. In general, younger alluvial fans are Holocene age, intermediate are late Pleistocene, and older alluvial fans are mid-Pleistocene or older. These age assignments are based on relationships with Lake Bonneville shoreline levels and correlation with similar units in California and Arizona dated by soil profile development (Shlemon & Purcell 1976) and radiometric age determinations (Bull 1974). Examples of younger and intermediate age fans are shown in Figure 1.

As is apparent from Tables 2 and 3, mapping of fans of various ages is important in evaluating terrain and surface cementation. Degree of dissection of alluvial fan deposits increases with age, as does the development of the surface soil profile. Older deposits exposed to weathering characteristically develop caliche in lower horizons of the soil profile. Degree of caliche cementation increases with age (Gile 1966) and caliche generally occurs only in intermediate and older alluvial fan deposits (Table 2).

## ENGINEERING EVALUATION OF GEOLOGIC UNITS

Selected physical properties and characteristics of basin-fill deposits were identified to provide data for making engineering judgments on road design and construction, shelter excavation, and to locate sources of aggregate. The properties selected for study are those

**Table 3** Photogeologic criteria for determination of relative age of alluvial fan deposits

| Relative age of alluvial fan | Drainage pattern | Drainage texture | Depth of stream incision | Topographic position above active channels | Landform | Pavement/ patina development |
|---|---|---|---|---|---|---|
| Younger | Distributary | Fine | None–low | Low | Undissected fan surface | None–poor |
| Intermediate | Dendritic | Medium | Low–moderate | Moderate | Dissected fan surface | Poor–well |
| Older | Dendritic | Coarse | Moderate–high | High | Highly dissected with original fan surface removed | None (eroded) |

**Table 4** Physical and engineering properties of geologic units

| Geologic unit | Grain size[a] | | | | | USCS symbol(s)[b] | Plasticity of fines | Consistency | Induration | Depth to indurated layer (cm) |
| | Max. size (mm) | Percent > 3 in. (76 mm) | Gravel | Percent < 3 in. (76 mm) Sand | Fines | | | | | |
|---|---|---|---|---|---|---|---|---|---|---|
| A1f | 22 | 0 | 0 | 9 | 91 | ML | Medium | Stiff | None | — |
| A1s | 45 | 0 | 3 | 79 | 18 | SM | None | Loose | None | — |
| A2s | 64 | 0 | 9 | 76 | 15 | SM | None | Loose | None | — |
| A3s | 25 | 0 | 0 | 90 | 10 | SP, SM | None | Medium dense | None | — |
| A3d | 7 | 0 | 0 | 95 | 5 | SP, SM | None | Loose | None | — |
| A4f | 3 | 0 | 0 | 6 | 94 | ML, CL | Low | Stiff | None | — |
| A4of | 10 | 0 | 0 | 9 | 91 | ML, CL | Low | Stiff | None | — |
| A4os | 74 | 1 | 10 | 76 | 14 | SM | None | Medium dense | None | — |
| A4og | 80 | 2 | 69 | 29 | 2 | GP | None | Loose | None | — |
| A5yf | 23 | 0 | 1 | 25 | 74 | ML | None–low | Stiff | None | — |
| A5ys | 78 | 0 | 12 | 73 | 15 | SM | None | Medium dense | None | — |
| A5yg | 150 | 6 | 57 | 33 | 10 | GM, GP | None | Medium dense | None | — |
| A5is | 105 | 1 | 16 | 67 | 17 | SM | None | Medium dense | Weak | 31 |
| A5ig | 146 | 6 | 58 | 30 | 12 | GM, GP | None | Medium dense | Weak | 29 |
| A5os | 126 | 2 | 14 | 71 | 15 | SM | Low | Medium dense | Moderate | 25 |
| A5og | 226 | 9 | 61 | 30 | 9 | GM, GP | None | Dense | Strong | 13 |

[a] Grain size subdivisions in the Unified Soil Classification System are defined in English units. These are retained with metric equivalents given parenthetically.
[b] Predominant USCS symbol; unit may consist of other soil types in appreciable amounts.

which could be easily determined in the field, and yield meaningful data for engineering evaluations. The most important of these characteristics are grain size distribution and plasticity of fines. Additional properties recorded include consistency, induration, and thickness of loose surface material. All properties were described using American Society for Testing and Materials (ASTM 1976) procedures adapted for field identification of soil characteristics. Data on physical properties of various geologic units are presented in Table 4. Data depicted in this table were derived from over 1250 stations established in twenty separate valleys in the Basin and Range Province.

## Road construction

Road construction considerations consist of evaluation of subgrade conditions, road base, flooding potential, and drainage crossing design. Favorable subgrade conditions require minimum surface preparation or improvement to the native soil. Desirable qualities are presence of granular materials of relatively high density, absence of large percentages (less than 10 to 15%) of loose or low-density fine-grained materials and absence of soils that exhibit high shrink–swell potential. A review of Table 4 shows that the fine-grained fluvial, lacustrine, and alluvial fan units are poor subgrade materials requiring several feet of excavation and recompaction. Depositional environments of high energy such as coarse-grained alluvial fan and lacustrine shoreline deposits provide fair to good subgrade due to the high percentage of mixed granular materials, necessitating minimal surface preparation.

Road base is recompacted select material overlying the subgrade and forming the wearing surface. Good road base will consist of well-sorted coarse-grained material with a low percentage of fines. Coarse-grained alluvial fan deposits and older lacustrine shoreline deposits are good to excellent as road base. Table 4 shows that in general all other units have smaller percentages of granular material and are not suitable for use as road base. Eolian and fluvial deposits with high percentages of well-sorted sand have both poor subgrade and road-base qualities. This is due to the great thickness of loose sand, absence of coarse material, and absence of fine-grained material which increases the compaction and moisture retention characteristics of a soil.

Flooding potential and number of drainage crossings are additional characteristics of road design and construction. High flooding potential would exist in those landforms with poorly drained or active surfaces of transport and deposition. Playas, young alluvial fans, and modern streams are examples. Drainage crossings would be minimal in these landforms; however, other measures would have to be taken to protect the road. Abandoned surfaces of material transport such as older and intermediate alluvial fans and stream terraces have low flooding potential. However, since active material and fluid transport is confined to numerous incised drainages, many drainage crossings are required.

## Excavation

Excavation conditions are important in the siting of critical facilities requiring partial burial. The critical excavation factors are ease of excavation and stability of excavation walls. Both factors vary with induration, consistency, and grain size, but optimum conditions for ease of excavation are generally opposite those needed to maintain stability of excavation walls. Easily excavated materials are those which are loose, uncemented, and lacking in coarse clasts. Nearly all materials composed predominantly of sand or fines are easily excavated. Sandy intermediate and older alluvial fan deposits may present a potential problem for excavation if near-surface caliche is indurated. All gravelly deposits are potentially difficult to excavate due to the presence of large boulders and cobbles. Gravelly alluvial fan deposits present the greatest problem due to the wide range of grain sizes and large maximum particle size (Table 4). The more well-sorted lacustrine gravels present few problems because they do not contain these large clasts and generally lack cementation.

Stability of vertical excavation walls is dependent on the percentage of fine-grained binding material, consistency, and induration. Eolian sands and lacustrine or younger alluvial fan gravels are subject to caving due to lack of clay binder or induration and loose consistency. Fine-grained, stiff soils of modern playa, older lacustrine, or alluvial fan deposits are very stable, with sandy lacustrine and alluvial fan deposits exhibiting low to moderate stability due to the presence of fines and medium-dense consistency. Medium dense to dense cemented intermediate and older alluvial fan deposits, whether sandy or gravelly, are moderately stable in excavation walls.

Overall excavation suitability represents a compromise due to the conflicting requirements for ease of excavation and wall stability. In general, excavation suitability is independent of depositional environment and varies chiefly with grain size and induration of deposits. Fine-grained soils are most suitable, sandy soils containing fines are intermediate in terms of suitability, and loose, granular sandy and gravelly soils are least suitable except when slightly indurated.

## Concrete aggregate

This discussion addresses only the aspects of aggregate suitability which relate generically to types of basin-fill deposits. Such aspects as the presence of reactive materials and durability which vary with local rock type are not addressed. Both fine and coarse aggregate will be required for construction. The deposits with the greatest potential for use as concrete aggregate are poorly sorted and contain few fines, reducing the need for processing. Gravelly older lacustrine deposits, particularly shoreline deposits of Lake Bonneville, contain few fines and are a good source of coarse aggregate unless too well sorted. Gravelly alluvial fan deposits contain a higher percentage of fines, and are locally cemented, but contain a wider range of grain sizes than reworked lacustrine deposits and are also suitable. Eolian sands, although free of fines, make poor fine aggregate due to well-rounded grains, lack of gradation (well sorted), and small grain size. Sandy fluvial, lacustrine, and alluvial fan deposits are composed of more angular sands, but also generally contain a higher percentage of fines and require washing. All fine-grained deposits (silts and clays) are unsuitable for use as concrete aggregate.

## DISCUSSION

The following discussion applies the methodology to an evaluation of specific engineering problems and presents an assessment of construction conditions in Spring Valley in Nevada. Spring Valley is in eastern Nevada, approximately 80 km east of Ely. Figure 2 is a portion of our surficial geologic map of Spring Valley and shows the wide variety of basin-fill deposits in the valley. The map illustrates the complexity of mapping, number and distribution of geologic and engineering data stops, and the distribution of basin-fill

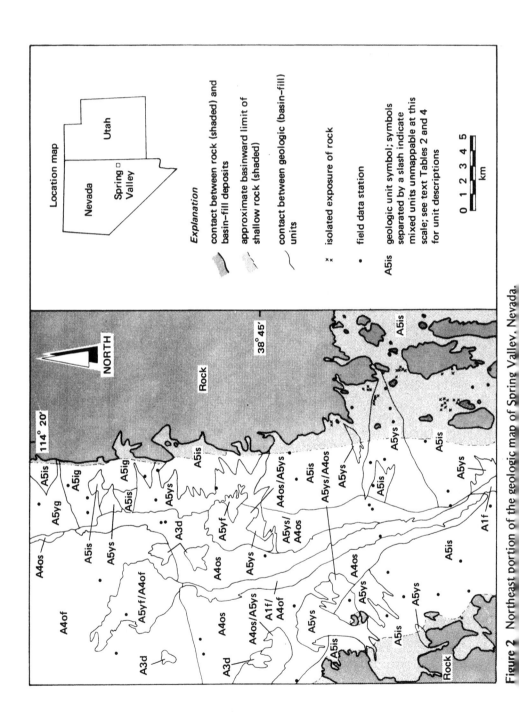

Figure 2  Northeast portion of the geologic map of Spring Valley, Nevada.

104

deposits within the valley. Spring Valley trends northwesterly, bounded to the east by the Snake Range, which includes Wheeler Peak, one of the highest peaks in Nevada, and to the west by the relatively low Schell Creek Range. The principal rock type in the area is limestone, with rhyolite, welded tuff, and basic volcanics present in the southern part of the valley. Areas of shallow rock and pediments fringe the valley as indicated by numerous isolated rock outcroppings and low-lying volcanic ridges around the periphery (Fig. 2).

Landforms and deposits in the valley are typical of those found in most Basin and Range valleys. Alluvial fans of various ages and grain sizes as well as fluvial, eolian, and lacustrine deposits occur in the valley. Alluvial fan deposits are gravelly and steep (locally >10%) on the east side. The materials here are deposited by streams from the Snake Range characterized by relatively large drainage basins with high relief. Alluvial fans on the west side of the valley are much more gentle and sandy. Deposits related to a Pleistocene lake in the valley occur in the northern part (Fig. 2). Shoreline deposits mapped in Spring Valley are sandy, not gravelly, since this was a relatively small lake fed predominately by streams carrying sand. Because of the small size and shallow depth of the lake, little wave action was present and concentrations of reworked gravels indicative of high-energy beaches common to Lake Bonneville do not occur. Eolian sands are present along the east side of the valley at the old shoreline level. Intermediate alluvial fans occur above the highest lake shoreline, with younger alluvial fans generally occurring below the shoreline, where shoreline bars have been breached and deposits have been reworked. Older lake basin silts and clays, as well as minor modern playa deposits, also occur below the shoreline bars.

Data on physical properties of these units were taken from over 100 locations in the valley (Fig. 2). These data were compiled into specific unit descriptions similar to the generalized descriptions in Tables 2 and 4. Variations in the Spring Valley data from the generalized description were slight. In general, the consistency of the deposits of Spring Valley was less dense than similar deposits occurring throughout the Basin and Range Province. Maximum particle size tended to be slightly smaller, and percentage of fines slightly larger than for generalized unit characteristics.

Using these data, the physical properties of each geologic unit within the valley were evaluated against the guidelines defining acceptable or unacceptable road, excavation, and aggregate parameters. Table 5 presents the results of this evaluation. Although they are general in nature and define average characteristics throughout the valley, they alert the planner or design engineer to conditions that may differ from typical or average facility design. Geologic hazards, problem soil, and difficult construction conditions are identified together with their areal extent. Having this information in advance of actual facility siting will allow flexibility in the final selection. More detailed surveys can now be focused which will define specific characteristics in a cost-effective manner.

## CONCLUSIONS

The methodology and example presented in this paper show how geomorphic analyses that combine interpretation of aerial photography with collection of field data on simple index properties can benefit the planner and design engineer in large regional siting projects. Problem soils, geologic hazards, and difficult construction conditions that will affect final cost and design can be identified early in the siting process. This approach can

# Table 5 Construction conditions, Spring Valley, Nevada

| Geologic unit | USCS symbol[a] | Subgrade | Road base | Flooding potential | Drainage crossings | Excavatability | Stability of vertical cuts | Concrete aggregate source |
|---|---|---|---|---|---|---|---|---|
| A1f | ML | Poor | Not suitable | High | Required | Easy | Stable | Not suitable |
| A2s | SM | Fair to good | Poor to fair | Low | None | Easy | Moderately stable | Poor |
| A3d | SP | Fair | Poor to fair | None | None | Easy | Unstable | Poor |
| A4of | ML, CL | Poor | Not suitable | High | None | Easy | Moderately stable | Not suitable |
| A4os | SM | Poor to good | Poor to fair | High | None | Easy | Moderately stable | Poor |
| A5yf | ML | Poor | Not suitable | Medium to high | Required | Easy | Stable | Not suitable |
| A5ys | SM | Poor to good | Poor to fair | Medium to high | Required | Easy. | Moderately stable | Poor |
| A5yg | GM, GP | Good | Good | Medium | Required | Moderately easy | Moderately stable | Good |
| A5is | SM | Fair to good | Fair to good | Low | Required | Moderately difficult to easy | Moderately stable | Poor |
| A5ig | GM, GP | Good | Good | Medium | Required | Easy | Moderately stable | Good |

[a]Predominant USCS symbol; unit may contain other soil types in appreciable amounts.

be applied to any siting program which requires rapid and inexpensive evaluation of construction conditions over large areas with conditions of Basin and Range type. Such projects as pipelines, transmission lines, highways, or critical facility siting would have other specific construction or design sensitivities such as potential for liquefaction, collapsible soils, or dating of faults. However, each could be evaluated regionally using an approach similar to that illustrated in this paper.

## REFERENCES

American Society for Testing and Materials 1976. *Annual book of ASTM standards*, Part 19, Procedure D-2488-69, Description of Soils (Visual–Manual Procedure), 314-20.

Bull, W.B. 1974. Geomorphic tectonic analysis of the Vidal region, Vidal Nuclear Generating Station, Information Concerning Site Characteristics, App. 2.5-B.

Bull, W.B. 1978. The alluvial fan environment. *Prog. Phys. Geography* 1, 222-70.

Gile, L.H. 1966. Morphological and genetic sequences of carbonate accumulation in desert soils. *Soil Sci.* **101**, 347-60.

Shlemon, R.J. and C.W. Purcell, 1976. Geomorphic reconnaissance, southeastern Mojave Desert, California and Arizona. Early Site Review Rept., Sundesert Nuclear Plant PSAR, App. 2.5-M.

# 8

# CRITERIA FOR CONSTRUCTING OPTIMAL DIGITAL TERRAIN MODELS

*Richard G. Craig*

## ABSTRACT

Digital terrain models (DTMs) have applications in diverse fields of geomorphology, including land use inventories, land reclamation, engineering, floodplain zoning, teaching, and military uses. The objective of the model builder is to capture the variability of the landform while minimizing the extraneous data collection. Unfortunately, to date, no explicit criteria have been developed that allow the researcher to plan a sampling scheme that is optimal for his or her particular purposes. Such criteria should be specific for the area under investigation and thus will involve a feedback between the data and the sampling procedure.

In this paper, a criterion is suggested that measures the quality of the DTM, and a method is presented that will produce the best DTM for chosen levels of the criterion in any terrain. It is based on a measure of the degree of independence of the individual observations comprising the DTM given by the autocorrelation function (ACF). It is shown that the ACF of landforms typically displays an exponential decay at increasing sample spacings, converging to an ACF characteristic of independent observations.

## INTRODUCTION

A topographic map is a collection of squiggly lines constructed according to definite rules. As such, it is an analog representation of the land surface. Its success is attested by the volume of sales, exceeding 6·6 million annually (U.S. Geological Survey 1980). Beginning in the 1940s, topographic map production increasingly made use of aerial photographs, replacing tedious field surveys. In the 1950s, the floating dot method was introduced; later, stereoscopic viewing became semiautomated through the addition of a computer, making a hybrid system. Now, with the advent of microcomputer technology the production process is fully automated, using analytical stereoplotters, and requiring only checking by an operator. This has cut the completion time for the typical map roughly in half (Teicholz & Dorfman 1976) and a single 7½ minute map can be produced within about 10 minutes (Scarano & Brumm 1976). This miracle is achieved through an intermediate step, which is the production of a digital data file of elevations, called the digital terrain model (DTM): the digital equivalent of the analog map. Applied scientists and engineers are now finding that the DTM is, in many cases, more convenient than the topographic map for problem solving.

Large numbers of people in diverse professions already use the numerical representation of maps in preference to the more conventional means. We need not search too far to find examples of such uses. In civil engineering, right-of-way design now routinely involves direct numerical computations from special terrain data bases, saving weeks of laborious hand calculations (Love 1972; Ternryd 1972; Turner 1978). Wherever earth moving is required, these models are used; examples include strip-mining operations (to compute volume of overburden), building excavations, channel straightening, and volume calculations for proposed reservoirs. Other uses include line-of-sight studies for microwave relay towers, rights-of-way for transmission lines, and planning forest fire control (Gossard 1978). Such uses can be expected to expand rapidly in the near future (Doyle 1978).

Another area currently depending heavily upon the DTM is remote sensing, especially that involving LANDSAT imagery. Here imagery is registered pixel by pixel to elevation information (Strahler & Estes 1980). For example, it has been found that moose and other wildlife habitat in Alaska can be mapped more accurately with this combination of data than with either method alone. Similarly, crop yield predictions such as the LACIE program are improved using such systems of information.

Geographic information systems are now growing in popularity and the DTM is an integral part of many (Smith & Blackwell 1980). A more pedestrian application of the DTM arises because three-dimensional models are easily constructed from such data (Fig. 1). Thus they are useful in promotional campaigns and wherever clear visual presentations are required, such as landscape architecture (Peucker 1972). Such presentations are also quite useful in geomorphic teaching.

More exotically, a DTM forms the heart of the guidance system for the proposed cruise missile. The missile must sense the landform, convert it to digital form, search its "remembered landform" for a match, then compute position and course corrections from those data as needed. The DTM is not just for fun!

The preceding sketch may provide an intuitive grasp of what is meant by a DTM; however, our intuition must be refined by rigorous geomorphic reasoning if we are to be sure that the DTM is a valid representation of the landform. Because the geomorphic phenomenon, the landform, is now expressed numerically, we must similarly expect our geomorphic reasoning to be translated into the language of numbers–mathematics–to

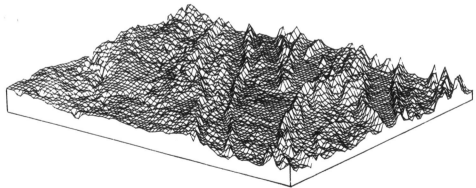

**Figure 1** Example of the application of a digital terrain model: the production of three-dimensional perspective views.

ensure that the "best" model is achieved. Accordingly, our first step is to state the pertinent geomorphic facts and relationships. Next, we convert these to precise mathematical terms. Finally, we employ the mathematical equivalence to identify the criteria of optimality. The procedure will be illustrated by several examples, and some interesting implications of the exercise will be shown.

## COMMON NOTIONS:  WHAT GEOMORPHOLOGISTS NEED FROM A MAP

Any general method of representing the landform must be capable of representing every "significant" landform. A number of methodologies have been presented which are applicable to specific landforms, for example the equations of Troeh (1965). Although of use for the particular problem studied, they are not suggested as a general method of representation. Any method that is universal must allow representation of any geomorphically significant landform. A suitable operational definition of *geomorphically significant* might be any landform that is mentioned in the *Encyclopedia of Geomorphology* (Fairbridge 1968).

The accuracy of the present 7½-minute maps can be taken as a standard for any representation. The general landform representation must be capable of attaining such accuracy as a minimum, and we must be able to represent the 7½-minute quad as conveniently as at present. Such maps, at a scale of 1:24 000 or 1:25 000, should be representable with reasonable effort. Most of us work at that scale or at smaller scales and most "classical" geomorphology was done at considerably smaller scales.

To put the question of simplicity of representation in a proper perspective, consider how much data is contained in a map that satisfies the national map accuracy standards. These standards require that no more than 10% of well-defined test points have a horizontal error greater than 1/50 in. or 40 ft at a scale of 1:24 000 (U.S. Geological Survey 1970). If errors are normally distributed with this range, the standard deviation of error would be 28 minutes. If one sample were taken at every 28 minutes, then to cover the land surface of the earth would require about $1.6 \times 10^{12}$ samples. By way of comparison, the most sophisticated computer storage systems *presently contemplated* have no more than $1 \times 10^{11}$ storage spaces! Thus a system beyond the present frontiers of our abilities would be required (Bell 1980).

It is therefore certain that no geomorphology that is both rational and complete can require such enormous capabilities. We know that not all of the forms that we see are worth portraying. There is value in generalizing, and such generalization involves a trade-off between precision and time. The best precision would require infinitely long to achieve; in effect, there is a macro-scale uncertainty principle in operation: the product of "speed" and "precision" can never exceed a certain critical constant. Of the precision, geomorphologists will recognize a certain portion of the variability of the landform as "noise" that it is not essential to study in order to reach a reasonable model of geomorphic processes. The remainder is the landform, that which must be understood, and it is the variability of this part which the geomorphologist hopes to capture. This is also the portion subject to trade-offs; there is no question that the noise can and should be sacrificed. An example may clarify the point.

Savigear (1965) has argued that, for the purposes of morphological mapping, we can think of a landform as consisting of individual rectilinear facets that can be classified as members of a small variety of possible segments. Neither Savigear, nor those who use his procedure, would suggest that this notion is absolutely correct.

Any area will differ at nearly all points from the assigned morphologic form. However, this variability is small compared to the degree of general agreement. Thus we may think of the variability explained by the form and contrast it to that which remains unexplained, the variability *from* the form, usually called the error.

To the extent that any systematic patterns remain in the variability from the form, we have an incomplete model. We strive to avoid such models and logically, we prefer models and techniques that enhance detection of their own failures.

Another feature that we require of our maps is that they be physically usable. No geomorphologist will agree to work with a set of maps that requires a storeroom to contain and which cannot be arranged conveniently for analysis on a moderate-size table. Thus we require that storage costs be small. Related to the size problem is the question of analysis time. We do not want our map to be presented in such an obscure or pedantic fashion as to waste analysis time by the interpretation burden. The maps should be clear and concise.

Let us summarize the common notions as they now appear:

(a) Every significant landform should be representable.

(b) The volume of data should be minimum.

(c) The minimum size that must always be representable is about 8 m.

(d) Any method must make explicit the trade-off between precision and time. Thus measures of precision should be available as part of the system.

(e) The method of representation should identify and remove the noise.

(f) It should provide signals when and if it fails.

## CONVERTING NOTIONS TO METHODS

These common notions form the basic considerations that must be satisfied, but they do not *generate* the landform (Aho & Ullman 1968). We still require the rule(s) for this. We first note that two questions occur when one attempts to quantify a landform. First, how many numbers are needed? Second, at what points should the numbers be obtained? The answers to these questions are closely related and define the appropriate DTM.

Concerning the first question, it should be clear that we do not need infinitely many samples, or else how could we expect our brain to do the analysis? Concerning the second question, if you can say precisely where you need them, that tells you how many samples are required. But we also want the information uniformly dispersed; that is, we do not want "holes" in the data matrix. Thus some regular space-filling polygon must be used. The available space-filling polygons are the regular square, the regular triangle, and the regular hexagon (Haggett & Chorley 1969, p. 51). The latter has the disadvantage that it is less convenient to store. The slight information-conveying advantage of the hexagonal grid does not offset the ease of use of the square grid (see, e.g., Singer & Wickman 1969). Thus we usually prefer to use the square grid. Therefore, the questions of interest to us are: (a) What is the appropriate distance between points? and (b) What orientation of grid should be used? Note that there is no advantage and considerable increase in complexity if we adjust the orientation to a particular perceived structural pattern or topographic "grain" since we either eventually get off of the pattern for which adjustment was made, or we must change orientation and thus lose the space-filling properties! We can therefore assume a grid oriented north–south and east–west. Thus the only pertinent question concerns the distance between points. There is no a priori justification for a different spacing in the north–south as opposed to the east–west direction, but the possible need for different spacings should be investigated.

To understand the other pertinent geomorphic constraints on landform representation better consider Figure 2. A regular grid placed parallel to the structure in any part of this section can never represent 100% of the form given in the figure. No discrete set of data can ever represent a continuous (and continuously differentiable) function. But no geologist can ever represent the landform to 100% accuracy in any other way either. And no geologist would argue that we must know every minute detail of every gulley or ravine before we can succeed in describing the essential geomorphic features of that terrain. Thus the question is not: Is the digital representation valid? Rather, we need to know: How do we construct a DTM at the level of accuracy that we require? This, it would seem, rests upon our decision on what proportion of the variability of the landform requires representation. Let us examine more closely the notion of variance of the landform.

An area $A$ is chosen arbitrarily and its boundaries defined. We assume that the landform surface is continuous or has no more than a finite number of discontinuities (cliffs). Then the mean elevation of that surface is well defined and is

$$\mu_z = \int_A z f(z) \, dz \tag{1}$$

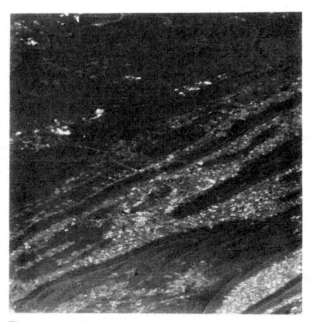

**Figure 2**  LANDSAT view of the Appalachian mountains in central Pennsylvania.

Similarly, the variance of elevations, one measure of the ruggedness of the terrain, is given by

$$\sigma_z^2 = \int_A z^2 f(z) \, dz - \mu_z^2 \qquad (2)$$

Such population parameters are not available to us; we must estimate them from finite samples. To the extent that our samples are random our estimates will be unbiased.

If the surface level within the area is uniform except for random fluctuations (which we expect to follow the normal distribution), the surface is completely described by these parameters. Thus we assume that we can divide completely the landform of interest into nonoverlapping regions within each of which the geomorphic conditions are uniform and the elevations are determined by random fluctuations about some "typical" level.

These assumptions, however, lack some intuitive appeal on one point: namely, if the elevations are assumed constant within such areas, the boundaries of each area are points of discontinuity and the entire region takes on a discontinuous, stair-step-like character not typical of real landforms on any scale presently observable (Fig. 3). How-ever, there is a large body of geomorphic observation and theory suggesting that *slopes* do obey this assumption (Savigear 1965; Gregory & Brown 1966; Doornkamp & King 1971). In addition, evidence presented below suggests that slope (defined as the differ-ence between successive values of elevation) has more desirable analytic properties. Mendal (1977) has suggested that observations of elevation constitute a self-similar process and, if so, they are inherently unsuitable for standard analytical methods.

When measured at the finest scales, adjacent slopes differ little from one another in value. Indeed, the differences are usually so slight that they are smaller than one would expect from random samples chosen on that slope facet. As we increase the distance

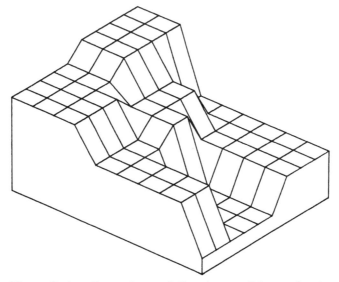

**Figure 3** Landform that satisfies the conditions of uni-formity. Within each subregion the geomorphic conditions are uniform.

**(a)**

**(b)**

**Figure 4**   Diverse landforms illustrating the relation of adjacent slopes: (a) Arizona, near the south rim of the Grand Canyon; (b) central Pennsylvania.

between samples of slope beyond the minimum practical spacing, we expect that the relation between such slope segments will decrease. This intuitive notion of landform behavior is a guide to check on the mathematical analysis to follow.

Let us examine this notion in a few examples. In Figure 4 are shown several diverse landforms. In each, the slope of the land surface near the photographer is quite uniform. Invariably as one moves farther away, new and distinct values of slope arise and even slopes of opposite sign become common.

Eventually, that is, at a sufficiently large distance, the slopes again regain their "original" value. However, these are separated by sufficiently distinct slope values that this must be counted as a new occurrence and is thus a distinct slope facet. Thus we can generalize our notion of landforms as consisting of a repetitious sequence of slope facets whose values are related to those of adjacent slope facets and are less closely related to more distant facets. Tobler's first law of geography thus apparently applies to landforms. This relation mirrors the inherent assumptions of the governing equation of slope development, the diffusion equation:

$$\frac{\partial z}{\partial t} = a \frac{\partial^2 z}{\partial x^2} + b \frac{\partial z}{\partial x} \tag{3}$$

where $z$ = elevation
$\quad\quad t$ = time
$\quad\quad x$ = distance from divide
$\quad\quad a, b$ = constants

This obviously does not rule out the fact that several facets in a region may have nearly identical values, but this is quite different from the intimate relation existing between adjacent slopes in which the values of one are determined, at least in part, by the values of another. Rather, in the case of the more distant, distinct facets with nearly identical values it is more likely that this similarity arises from other geomorphic factors, such as similar lithologic foundation, climatic, and/or structural conditions.

Indeed, we may find that a suitable definition of a slope facet is a subregion of topography within which the slope value is significantly related to adjacent values. And this appears to agree with our intuitive notions and previously advanced definitions.

It remains for us to see how this definition of facet can help us solve our original problem, the grid spacing of a DTM.

Let us recap our available constraints:

1. The DTM consists of discrete data points.
2. The points are arrayed in a rectangular grid.
3. The grid axes are fixed north–south and east–west.
4. The grid spacing will be uniform throughout the area.
5. The grid spacing will be identical north–south and east–west unless strong reasons otherwise are given.
6. The data represent elevation at the point in question.
7. The spacing is to be based upon characteristics of differences in elevation (slope), namely that the spacing is just large enough so that the slopes are unrelated to one another.

With these ideas now clearly before us, it is easy to see that our criterion 7 must somehow reflect the typical spacing that will yield samples unrelated to one another. By typical we mean that we choose here some measure of the average size of the facets, and this most clearly shows us that the DTM will be particular to a single physiographic region since we can expect the average to differ from region to region (almost by definition). But how can we measure the relatedness of adjacent facets in an objective fashion that will allow us to know when the values of two facets are not related?

Clearly, the quesion of relatedness is a statistical one, and a typical statistical test of relatedness is the correlation coefficient. Thus a simple procedure would be to divide the area into very small facets. Since the simplest way to define the facet requires three points, we will term the method *triangularization*. For each triangle the inclination may be computed. A sufficiently large sample of pairs of facets will allow us to compute the correlation coefficient; this coefficient can be tested for significance (is it nonzero?). The process is then repeated for larger and larger facets until a nonsignificant coefficient is obtained. In practice, this is difficult to do.

Several problems arise in this procedure. First, because we are dealing with inclination, an angular measure, the regular correlation is inappropriate. Adjustments are available (Fisher 1953; Mardia 1972) but are not usually familiar.

Another problem is that the technique is somewhat unwieldy and exceedingly tedious. This makes the proper application of the procedure unlikely, except in a purely academic exercise. Even if the technique were followed, there is considerable difficulty in interpreting the results. How does one convert the size of a triangle into a spacing between square grid points? It would be less difficult if the grid were hexagonal, but this has been ruled out on other grounds. Thus the method can hardly be suggested as practical for the kinds of applications suggested in the introduction to this paper.

Fortunately, a second approach is available that offers considerable advantages to the first, and satisfies the previously defined criteria. Suppose that rather than triangulariza-

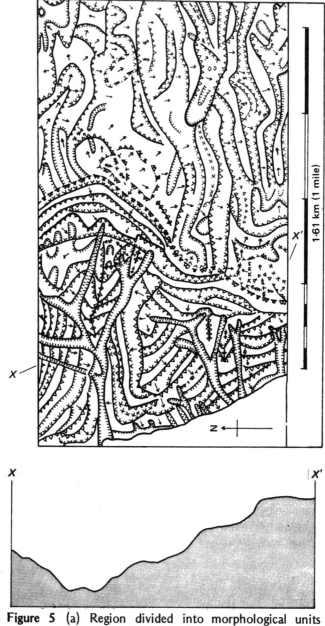

**Figure 5** (a) Region divided into morphological units according to the method of Savigear (1965). (b) Cross section constructed along a line *X-X'* randomly intersecting these units.

tion, we merely lay down traverses across an area. Although the traverse is arbitrarily placed it will, on the average, intercept the facets along a distance equal to their mean dimension (Kendall & Moran 1963). Along each traverse we observe elevations at regularly spaced intervals, as in Figure 5. A cross section constructed from these elevations is also shown. Note again that the slopes tend to approximate adjacent values quite closely and that the degree of approximation decreases as the distance between points increases. Thus evidently, we can use traverses equally as well to represent the "memory" in slopes and perhaps as a means of estimating the size of a typical slope. None of this information is particularly surprising, and if no more objective techniques were available, this would hardly be considered a worthwhile exercise; however, several quite useful and startling results follow directly.

Note that if we can estimate the number of sample points separating slopes which are "unrelated," we have the required estimate of grid size. A measure of relatedness for adjacent slopes is not hard to find; the traditional correlation coefficient will do quite handily. In this case we require the correlation between each slope and the next adjacent slope along the traverse. Such a correlation between a set of values and itself at some nearby points is termed the *autocorrelation*, and a vast literature is available concerning the properties of these autocorrelation coefficients (Box & Jenkins 1976). Some of these properties are quite important for our own analysis, as shown in the following section.

## SOME PROPERTIES OF THE AUTOCORRELATION COEFFICIENT

The classical correlation coefficient, termed the Pearson Product Moment correlation after its inventor, is defined by the following equation.

$$r_{xy} = \frac{\sum_{i=1}^{N} (x_i - \bar{x})(y_i - \bar{y})}{\sqrt{\sum_{i=1}^{N} (x_i - \bar{x})^2 \sum_{i=1}^{N} (y_i - \bar{y})^2}} \tag{4}$$

In this equation $\bar{x}$ and $\bar{y}$ represent the means of $N$ samples of the two variates $X$ and $Y$ whose correlation is to be calculated. In our case, the variates are the $S(I)$ (the slopes) and the slopes $K$ steps removed $S(I + K)$. Thus the *auto*correlation at lag $k$ is

$$r_k = C_k / C_0 \tag{5}$$

where $C_k$ is the autocovariance at lag $k$ defined by

$$C_k = \frac{1}{N} \sum_{i=1}^{N-k} (x_i - \bar{x})(x_{i+k} - \bar{x}) \tag{6}$$

The autocorrelation can be computed for up to $N - 1$ many lags, although in practice $k = N/4$ is the largest used. A plot of these values versus the lag is known as the auto-

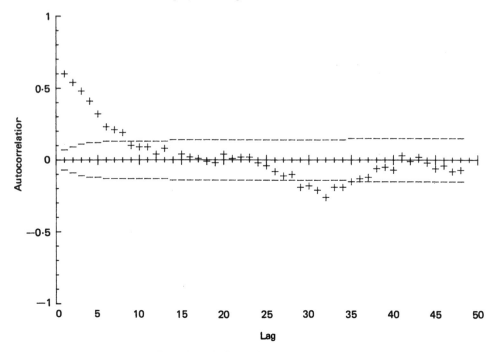

**Figure 6**   Autocorrelation function of slopes from an area in central Pennsylvania. Upper and lower one-standard-error limits are also plotted.

correlation function (ACF) and an ACF typical of slopes is shown in Figure 6. When the autocorrelation is computed for the traverse of elevations instead of the slopes, we get an ACF as shown in Figure 7. Notice that this plot decays quite slowly, starting at a value of approximately +1 and remaining positive throughout. Such behavior is typical of a *non-stationary* series, a series for which no mean exists. It is the nonstationarity of these elevation data that shows mathematically that the slopes (first differences of the elevation series) are more appropriate phenomena for study.

Returning to Figure 6, note that the value at lag 1, representing the mean autocorrelation between each slope and the next adjacent slope, is positive and high (0·60). Just as with the standard correlation coefficient, the autocorrelation can range between +1 and −1. The high positive value shows that, on average, each slope is followed by another slope very much like it. From the decay of the ACF we see that a slope two steps away from another is less likely to be the same, and so on until finally a point is reached at which the slopes are totally unrelated. Such behavior mirrors our intuitive notion of the way slopes should behave and provides a quantitative testable measure of the degree of relation between slopes. Characteristics of that test are described next.

As usual, we take as a null hypothesis that the correlation is zero. Then the observed value should be no more than about two standard deviations from zero (at the 5% confidence level), where the standard error is computed from the following expression:

$$\text{var}\,[r_k] = 1/N \left[ (1 + \phi^2)(1 - \phi^{2k})/(1 - \phi^2) - 2k\phi^{2k} \right] \tag{7}$$

The one-standard-error limits are plotted in Figures 6 and 7. Now we might look at the

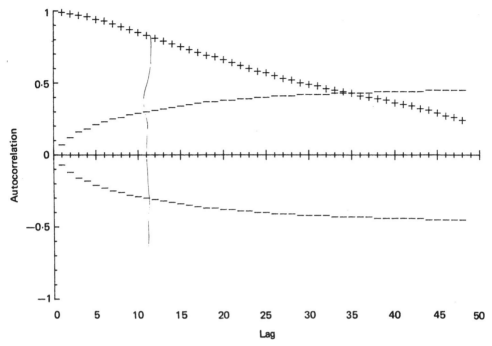

**Figure 7** Autocorrelation function of elevations along the same traverse as in Figure 6. Upper and lower standard error limits are also shown.

point where the ACF gets "close enough" to zero and take that lag as representing the distance required to yield unrelated elevations.

This would suggest $k = 36$ lags, which, since the sample spacing is 48 m, would imply a sampling interval of 48 m $\times$ 36 = 1728 m. If such a sample spacing (2 km) is used, it yields a landform as in Figure 8; this seems to be a useless representation (compare to Figure 1) and no one would believe that this captures the familiar physiography of the Appalachian mountains. This intuitive notion of "inappropriateness" is substantiated by

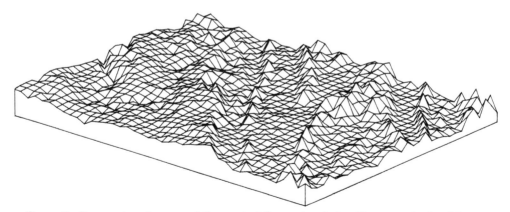

**Figure 8** Perspective diagram of the central Pennsylvania landform as obtained by sampling according to the decay of the ACF of elevations.

the statistical argument; the series is nonstationary, hence not amenable to analysis. The suggested cure from statisticians is to consider the first difference of the series, that is, the slopes.

If we use the decay of the ACF of slopes, a sample spacing of 432 m is suggested. But this can be improved. It is first important to note that these are estimates made from samples and so are subject to sampling fluctuations. If our traverse is $N$ samples long, $N - 1$ are used to estimate the ACF at lag 1, $N - 2$ at lag 2, and so on. Thus we have less certainty in the values of the ACF computed for the higher lags. The choice of lag giving an ACF "small enough" is subject to large uncertainty, as can be seen in Figures 6 and 7. However, we can use the value at lag 1 to estimate when the ACF is small enough because the theoretical ACF of landform traverses typically decays as

$$r_k = r_1^k \tag{8}$$

as shown earlier, and as anticipated from the diffusion equation. We would like to compute a value of lag at which the autocorrelation is not significantly different from zero. This, in turn, requires information on the sampling variance of $r_k$, which has been shown to be related to the chi-squared distribution (Bartlett 1946). The test is described in more detail elsewhere (Craig 1981), but suffice it to say that the equation of interest is

$$k \geq \ln\left[1 + n(1 - \phi^m)/X^2\right]/ - \ln(\phi^2) \tag{9}$$

where  $n$ = number of samples contemplated
  $m$ = number of lags included in the test
  $X^2$ = cutoff value of a chi-squared variate with $m - 1$ degrees of freedom at the desired confidence level
  $h$ = minimum sampling interval
  $\phi$ = value of the ACF at lag 1

## Example

A sample calculation may make the method more transparent. To provide information on the relation of slope to lithology, it was desired to collect a large number of independent observations of slope. The area of interest is central Pennsylvania. A grid pattern was chosen to facilitate computer analysis following the arguments presented earlier. Thirty-six quadrangles covering 5500 $km^2$ were to be sampled. A large portion of this section of the Appalachians was analyzed along a total of eleven traverses, each of length 200, at a very close spacing (48 m). The mean value of $\phi$ for these traverses is 0·98 and there are no significant differences from traverse to traverse (see Table 1). The number of samples anticipated is large, at least 1000. Only two lags will be tested, so $m = 2$. This is probably the best value to use. At the 95% level of confidence the cutoff for a chi-squared variate with one degree of freedom is 3·841. Using these values we compute

$$k \geq \ln\left[1 + 1000(1 - 0·98^2)/3·841\right]/ - \ln(0·98^2)$$

or

$$k \geq 60·03$$

**Table 1**  Values of $\phi$ for the Appalachians of central Pennsylvania estimated from eleven traverses, each of length 200, taken at a sample spacing of 48 m

| Traverse number | Lag 1 (autocorrelation = $\phi$) | |
|---|---|---|
| | Elevations | Slopes |
| 1 | 0·96 | 0·50 |
| 2 | 0·98 | 0·73 |
| 3 | 0·99 | 0·60 |
| 4 | 0·98 | 0·65 |
| 5 | 0·96 | 0·62 |
| 6 | 0·96 | 0·66 |
| 7 | 0·99 | 0·72 |
| 8 | 0·96 | 0·33 |
| 9 | 0·99 | 0·77 |
| 10 | 0·99 | 0·70 |
| 11 | 0·97 | 0·63 |
| Mean | 0·98 | 0·63 |
| Variance | 0·0001 | 0·01 |

Thus sampling at 60·03 × 48 m = 2881 m or roughly 3 km or more will yield unrelated (independent) samples of elevation. Sampling at this spacing produces a landform representation as shown in Figure 9. It is apparent that this spacing is too large to capture the familiar form of the Appalachians; again we suspect that the reason is that the elevation series is not appropriate for study; it is nonstationary. Instead, the *slopes* must form the basis of the decision on sample spacing.

The ACF of slopes computed from the same eleven traverses yield a mean value of $\phi$ of 0·63 ± 0·12. To ensure independence, we want a sample spacing that will provide independent samples in the worst likely case, when $\phi$ takes on a value indicated by the

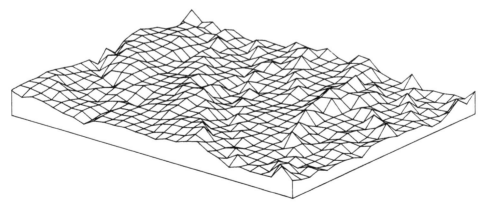

**Figure 9**  Perspective diagram of the central Pennsylvania landform as obtained from equation (9), but computed using elevations to estimate $\phi$.

**Table 2** Sample spacings computed by iteration until $N$ does not change[a]

| Number | $\phi$ | H | Final spacing | Samples required | Number of iterations |
|--------|--------|---|---------------|------------------|----------------------|
| 1 | 0·95 | 18 | 859·12 | 209 | 73 |
| 2 | 0·90 | 14 | 663·04 | 350 | 18 |
| 3 | 0·85 | 11 | 541·16 | 526 | 14 |
| 4 | 0·80 | 10 | 457·15 | 737 | 12 |
| 5 | 0·75 | 8 | 394·73 | 988 | 11 |
| 6 | 0·70 | 7 | 346·19 | 1285 | 11 |
| 7 | 0·65 | 6 | 306·89 | 1635 | 11 |
| 8 | 0·60 | 6 | 274·20 | 2048 | 10 |
| 9 | 0·55 | 5 | 246·32 | 2538 | 10 |
| 10 | 0·50 | 5 | 222·10 | 3122 | 10 |
| 11 | 0·45 | 4 | 200·72 | 3823 | 10 |
| 12 | 0·40 | 4 | 181·53 | 4673 | 10 |
| 13 | 0·35 | 3 | 164·06 | 5722 | 10 |
| 14 | 0·30 | 3 | 147·91 | 7039 | 10 |
| 15 | 0·25 | 3 | 132·72 | 8743 | 10 |
| 16 | 0·20 | 2 | 118·14 | 11034 | 10 |
| 17 | 0·15 | 2 | 103·74 | 14309 | 10 |
| 18 | 0·10 | 2 | 88·84 | 19513 | 10 |
| 19 | 0·05 | 1 | 71·76 | 29903 | 10 |

[a] These computations assume that one 7½-minute quad is to be sampled and that the value of phi was estimated from traverses with a sample spacing of 48 m.

mean plus two standard deviations, 0·87. This value will be exceeded in no more than 5% of the area. Using the value $\phi = 0·87$ we compute a sample spacing of 718 m. But if this were used we would have over 10 000 samples. Using this new value for $N$ in equation (9), we recompute $k$ until $N$ does not change. After twelve iterations we find $k = 1008$ m and, for simplicity, a 1-km spacing was finally chosen.

The computations can be programmed easily for a computer and iterations performed until $N$ does not change. In general, it has been found that the program will converge whenever $\phi \leq 0·95$. Typical results for various values of $\phi$ are given in Table 2.

Using this sample spacing, elevations were recorded for the Pennsylvania area and the results are depicted in Figure 1. As can be seen, the major features of Appalachian topography are adequately displayed. It is of some interest to note that Kirkby (1976, p. 12-13) has computed that one square kilometer is the minimum sustainable drainage basin area in a humid temperate climate. In addition, map units digitized at the same spacing accurately represent the stratigraphic sequence. That is, units are rarely skipped over, as can be seen in the geologic map in Figure 10. In fact, adjacent grid points contain stratigraphically adjacent units in 84% of the cases, as shown by the transition matrix in Table 3.

Of course, a technique such as this is only valid to the extent that the ACF consistently follows the exponential decay suggested by (8). To evaluate this consistency, a set of traverses from different areas of the country were collected and the ACF computed. These regions were chosen from a list of 100 maps illustrating diverse physiographic features (Upton 1955). Thirty-one of the forty areas that were listed as 7½-minute

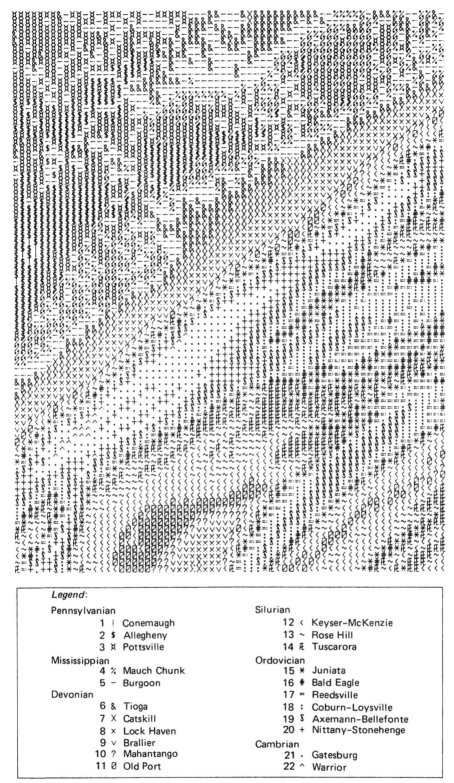

Legend:

| Pennsylvanian | | | Silurian | | |
|---|---|---|---|---|---|
| 1 | ! | Conemaugh | 12 | < | Keyser-McKenzie |
| 2 | $ | Allegheny | 13 | ~ | Rose Hill |
| 3 | ⋈ | Pottsville | 14 | ℞ | Tuscarora |
| Mississippian | | | Ordovician | | |
| 4 | % | Mauch Chunk | 15 | * | Juniata |
| 5 | − | Burgoon | 16 | # | Bald Eagle |
| Devonian | | | 17 | = | Reedsville |
| 6 | & | Tioga | 18 | : | Coburn-Loysville |
| 7 | X | Catskill | 19 | S | Axemann-Bellefonte |
| 8 | × | Lock Haven | 20 | + | Nittany-Stonehenge |
| 9 | v | Brallier | Cambrian | | |
| 10 | ? | Mahantango | 21 | · | Gatesburg |
| 11 | 0 | Old Port | 22 | ^ | Warrior |

Figure 10  Geologic map of the central Pennsylvania area obtained when the units are digitized at a 1-km spacing.

**Table 3** Matrix of transition frequencies between map units as displayed on the grid of Figure 10[a]

| | 1 | 2 | 3 | 4 | 5 | 6 | 7 | 8 | 9 | 10 | 11 | 12 | 13 | 14 | 15 | 16 | 17 | 18 | 19 | 20 | 21 | 22 |
|---|---|---|---|---|---|---|---|---|---|---|---|---|---|---|---|---|---|---|---|---|---|---|
| 1 | 2 | 21 | 0 | 0 | 0 | 0 | 0 | 0 | 0 | 0 | 0 | 0 | 0 | 0 | 0 | 0 | 0 | 0 | 0 | 0 | 0 | 0 |
| 2 | 21 | 898 | 298 | 45 | 97 | 7 | 0 | 0 | 0 | 0 | 0 | 0 | 0 | 0 | 0 | 0 | 0 | 0 | 0 | 0 | 0 | 0 |
| 3 | 0 | 298 | 1060 | 264 | 436 | 67 | 0 | 0 | 0 | 0 | 0 | 0 | 0 | 0 | 0 | 0 | 0 | 0 | 0 | 0 | 0 | 0 |
| 4 | 0 | 45 | 264 | 690 | 414 | 45 | 0 | 0 | 0 | 0 | 0 | 0 | 0 | 0 | 0 | 0 | 0 | 0 | 0 | 0 | 0 | 0 |
| 5 | 0 | 97 | 436 | 414 | 988 | 334 | 14 | 0 | 0 | 0 | 0 | 0 | 0 | 0 | 0 | 0 | 0 | 0 | 0 | 0 | 0 | 0 |
| 6 | 0 | 7 | 67 | 45 | 334 | 470 | 128 | 0 | 0 | 0 | 0 | 0 | 0 | 0 | 0 | 0 | 0 | 0 | 0 | 0 | 0 | 0 |
| 7 | 0 | 0 | 0 | 0 | 14 | 128 | 528 | 120 | 0 | 0 | 0 | 0 | 0 | 0 | 0 | 0 | 0 | 0 | 0 | 0 | 0 | 0 |
| 8 | 0 | 0 | 0 | 0 | 0 | 0 | 120 | 474 | 133 | 2 | 0 | 0 | 0 | 0 | 0 | 0 | 0 | 0 | 0 | 0 | 0 | 0 |
| 9 | 0 | 0 | 0 | 0 | 0 | 0 | 0 | 133 | 248 | 112 | 18 | 6 | 5 | 0 | 0 | 0 | 0 | 0 | 0 | 0 | 0 | 0 |
| 10 | 0 | 0 | 0 | 0 | 0 | 0 | 0 | 2 | 112 | 102 | 60 | 51 | 40 | 0 | 1 | 1 | 0 | 0 | 0 | 0 | 0 | 0 |
| 11 | 0 | 0 | 0 | 0 | 0 | 0 | 0 | 0 | 18 | 60 | 316 | 176 | 12 | 5 | 1 | 1 | 0 | 0 | 0 | 0 | 0 | 0 |
| 12 | 0 | 0 | 0 | 0 | 0 | 0 | 0 | 0 | 6 | 51 | 176 | 1012 | 255 | 76 | 29 | 1 | 1 | 0 | 0 | 0 | 0 | 0 |
| 13 | 0 | 0 | 0 | 0 | 0 | 0 | 0 | 0 | 5 | 40 | 12 | 255 | 410 | 263 | 156 | 21 | 8 | 0 | 0 | 0 | 0 | 0 |
| 14 | 0 | 0 | 0 | 0 | 0 | 0 | 0 | 0 | 0 | 0 | 5 | 76 | 263 | 310 | 180 | 72 | 23 | 2 | 0 | 0 | 0 | 0 |
| 15 | 0 | 0 | 0 | 0 | 0 | 0 | 0 | 0 | 0 | 1 | 1 | 29 | 156 | 180 | 172 | 141 | 197 | 62 | 19 | 0 | 0 | 0 |
| 16 | 0 | 0 | 0 | 0 | 0 | 0 | 0 | 0 | 0 | 1 | 1 | 1 | 21 | 72 | 141 | 78 | 130 | 104 | 8 | 1 | 0 | 0 |
| 17 | 0 | 0 | 0 | 0 | 0 | 0 | 0 | 0 | 0 | 0 | 0 | 1 | 8 | 23 | 197 | 130 | 242 | 303 | 60 | 2 | 0 | 0 |
| 18 | 0 | 0 | 0 | 0 | 0 | 0 | 0 | 0 | 0 | 0 | 0 | 0 | 0 | 2 | 62 | 104 | 303 | 303 | 476 | 320 | 36 | 2 |
| 19 | 0 | 0 | 0 | 0 | 0 | 0 | 0 | 0 | 0 | 0 | 0 | 0 | 0 | 0 | 19 | 8 | 60 | 476 | 560 | 186 | 49 | 6 |
| 20 | 0 | 0 | 0 | 0 | 0 | 0 | 0 | 0 | 0 | 0 | 0 | 0 | 0 | 0 | 0 | 1 | 2 | 320 | 186 | 352 | 226 | 9 |
| 21 | 0 | 0 | 0 | 0 | 0 | 0 | 0 | 0 | 0 | 0 | 0 | 0 | 0 | 0 | 0 | 0 | 0 | 36 | 49 | 226 | 660 | 31 |
| 22 | 0 | 0 | 0 | 0 | 0 | 0 | 0 | 0 | 0 | 0 | 0 | 0 | 0 | 0 | 0 | 0 | 0 | 2 | 6 | 9 | 31 | 24 |

[a]The numbers in the column and row headings correspond to those in the legend of Figure 10. The table is read as, for example, the number of times that unit 10 is adjacent to unit 9 is 112 (the entry in row 10, column 9).

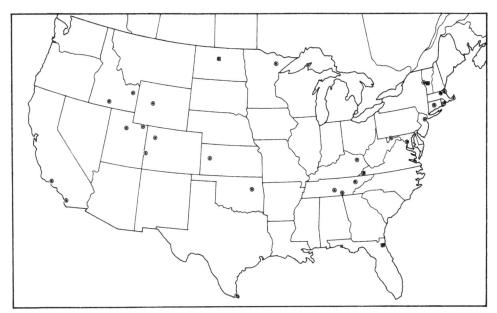

**Figure 11** Locations chosen from Upton (1955) to study autocorrelation properties of diverse landforms.

quadrangles were chosen (Fig. 11). On each map, four traverses each of fifty-one points spaced 2 mm (48 m on the ground) were recorded, one at each major compass heading, as shown in Figure 12. In all cases, except perfectly flat terrain, the ACF followed an exponential decay. Several typical ACFs are shown in Figure 13. The mean value of phi, the standard deviation, and the sample spacing as computed from (8) are given in Table 4. The values of phi within each site are, in general, remarkably uniform. The greatest variability comes where the terrain is flattest. There is some indication that the procedure may break down in areas such as tidal marshes, former lake bottoms, and other depositional-type terrains. In addition, the user should be cautioned that the traverses from which the sample spacing is estimated should *span* the entire terrain to be modeled. Shorter traverses may be inadequate, and for that reason the values reported in Table 4 may require revision as more extensive sampling is done. In general, these results tend to substantiate Southard's (n.d.) claim that up to "90% of the raw data collected can often be eliminated without imparing the visual quality or accuracy of the final 1:24 000 scale map data."

## CONCLUSIONS

We now have the capability of representing terrain in a computer compatible manner, but the available representations generally exceed our capability to analyze these data. Using a geomorphologically based criterion – namely the degree of interrelatedness of adjacent slope measurements – we can calculate a sample spacing that "captures" the essential features of the landform and is representable in a volume of data that are amenable to analysis.

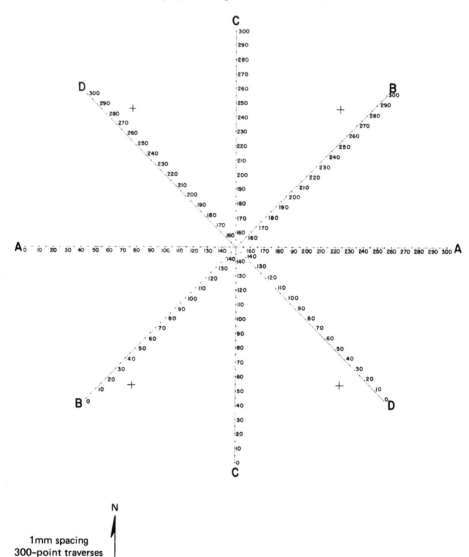

**Figure 12** Sampling procedure used to study the autocorrelation of diverse terrains. Elevations were recorded along each of the traverses shown and were estimated to the nearest contour interval.

As a rule, sample spacings of 100 m are more fine than is required to characterize any landform studied. Spacings up to 1 km may be adequate for a large number of landforms. Use of the method outlined here can relieve over 99% of the data collection, storage, and analysis burden imposed by the present forms of DTM.

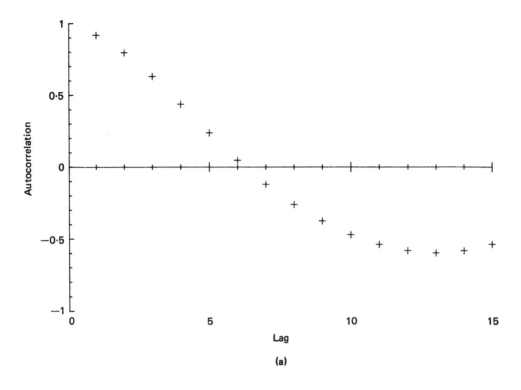

(a)

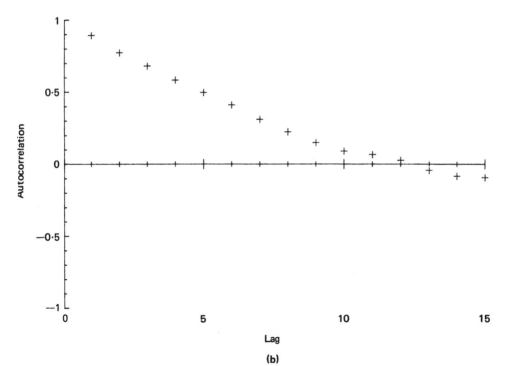

(b)

*(figure continues)*

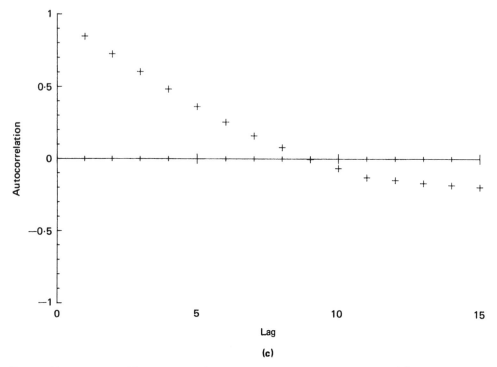

(c)

Figure 13   Sample ACFs recorded from several physiographic regions: (a) Ewing, Kentucky; (b) Delaware, Michigan; (c) East Brownsville, Texas.

Table 4   Autocorrelation properties and optimal sampling interval as determined by traverses of landforms in thirty-one quadrangles throughout the United States[a]

| Quadrangle name | Location | $\phi$ | $+/-^a$ | Spacing[b] | Landform |
|---|---|---|---|---|---|
| Ticonderoga | NY–VT | 0·95 | 0·02 | 847 | Adirondacks |
| Ewing | KY | 0·94 | 0·02 | 826 | Cumberland Mountains |
| Virginia | MN | 0·94 | 0·03 | 803 | Superior upland – metasediments |
| Juniata Arch | CO | 0·93 | 0·05 | 782 | Canyonlands |
| Flaming Gorge | UT | 0·93 | 0·03 | 777 | Middle Rocky Mountains |
| Menan Buttes | ID | 0·93 | 0·03 | 772 | Lava plain |
| Brandon | VT | 0·92 | 0·04 | 742 | Green Mountains |
| Cumberland | MD–PA | 0·92 | 0·02 | 719 | Valley and ridge – middle |
| Thousand Springs | ID | 0·92 | 0·07 | 713 | Snake River Plain |
| Whitwell | TN | 0·91 | 0·03 | 691 | Cumberland Plateau |
| Ayer | MA | 0·91 | 0·02 | 680 | Glaciated New England upland |
| Washington West | DC | 0·90 | 0·06 | 662 | Piedmont |
| Jacksonville Beach | FL | 0·90 | 0·03 | 656 | Barrier beach |
| Delaware | MI | 0·89 | 0·05 | 644 | Superior upland – sediments |
| San Luis Rey | CA | 0·89 | 0·05 | 622 | Arroyo |
| Ventura | CA | 0·88 | 0·03 | 618 | Alluvial fan |
| Lynn | MA | 0·87 | 0·04 | 593 | Seaboard lowlands |
| Jenks | OK | 0·86 | 0·09 | 571 | Central lowland |
| Kingston | RI | 0·86 | 0·08 | 566 | Glaciated seaboard lowlands |

**Table 4**  *(continued)*

| Hillsboro | KY | 0·85 | 0·02 | 537 | Lexington Plain |
|---|---|---|---|---|---|
| Rover | TN | 0·85 | 0·06 | 536 | Karst topography |
| East Brownsville | TX | 0·84 | 0·07 | 531 | Lower Rio Grande Valley |
| Maverick Spring | WY | 0·83 | 0·08 | 506 | Eroded dome |
| Anvil Points | CO | 0·79 | 0·03 | 820 | Colorado Plateau |
| Provincetown | MA | 0·78 | 0·10 | 430 | Seaboard lowland |
| Lake McBride | KS | 0·78 | 0·18 | 425 | Dissected high plains |
| Patterson | NJ | 0·74 | 0·22 | 383 | Glaciated piedmont lowland |
| Norris | TN | 0·72 | 0·19 | 368 | Strike Valley |
| Jordan Narrows | UT | 0·71 | 0·46 | 354 | Bonneville deposits |
| New Britain | CT | 0·67 | 0·46 | 318 | Glacial deposits |
| Voltaire | ND | 0·65 | 0·45 | 306 | West Lake section |

[a] The numbers in the fourth column represent the one-standard deviation limit.
[b] Sample spacings are based upon the maximum value of $\phi$ observed in the quadrangle assuming that $N = 154$ given in meters.

## ACKNOWLEDGMENTS

The following persons are to be thanked for help in collecting, keypunching, and computer analysis of the data used in this paper: James Warren, Daniel Duck, Steve Heimlich, Michael Raymondi, Scott Moyer, and Larry Wickstrom.

Ms. Dorothy Thompson provided expertise in computer typesetting to prepare the text of the manuscript.

This paper is Contribution No. 209, Department of Geology, Kent State University. Portions of this work were performed while the author was visiting faculty fellow at the Goddard Space Flight Center and were supported by NASA Grant NASA5-26111.

## REFERENCES

Aho, A.V. and J.D. Ullman, 1968. The theory of languages. *Math. Systems Theory* 2(2), 97-125.

Bartlett, M.S. 1946. On the theoretical specification of sampling properties of autocorrelated time series. *Jour. Royal Statist. Soc.* B8(27).

Bell, P. 1980. *EOS* 61(38), 628.

Box, G.E.P. and G.M. Jenkins 1976. *Time series analysis: forecasting and control*, 2nd edn. New York: Holden–Day, 553 p.

Craig, R.G. 1981. Sampling an autocorrelated process: the AR(1). *Jour. Internat. Assoc. Math. Geology* (in press).

Doornkamp, J. and C.A.M. King 1971. *Numerical analysis in geomorphology*. New York: St. Martin's Press, 372 p.

Doyle, F.J. 1978. Digital terrain models: an overview. *Photogrammetric Engineering and Remote Sensing* 44(12), 1481-5.

Fairbridge, R.W. 1968. *The encyclopedia of geomorphology*. Stroudsburg, Pa.: Dowden, Hutchinson & Ross, 1295 p.

Fisher, R.A. 1953. Dispersion on a sphere. *Proc. Royal Soc. London* A217, 295-305.

Gossard, T.W. 1978. Applications of DTM in the Forest Service. *Photogrammetric Engineering and Remote Sensing* 44(12), 1577-86.

Gregory, K.J. and E.H. Brown 1966. Data processing and the study of land form. *Zeitschr. Geomorphologie* NF10, 237–63.

Haggett, P. and R.J. Chorley 1969. *Network analysis in geography*. New York: St. Martin's Press, 348 p.

Kirkby, M.J. 1976. Deterministic continuous slope models. *Zeitschr. Geomorphologie* NF25, 2–19.

Kendall, and Moran 1963. Geometrical probability.

Love, R.S. 1972. The digital terrain model of the GLC ALICE system. *Proc. PTRC Seminar*, 1–7.

Mardia, K.V. 1972. *Statistics of directional data*. New York: Academic Press, 357 p.

Mendal, B.B. 1977. *Fractals form chance and dimension*. San Francisco: W.H. Freeman, 365 p.

Peucker, T.K. 1972. *Computer cartography*. Commission on College Geography, Resource Paper 17, Association of American Geographers, 75 p.

Savigear, R.A.G. 1965. A technique of morphological mapping. *Ann. Assoc. Am. Geographers* 35, 514–38.

Scarano, F.A. and G.A. Brumm 1976. A digital elevation data collection system. *Photogrammetric Engineering and Remote Sensing* 42(4), 489–96.

Singer, D.A. and F.E. Wickman 1969. *Probability tables for locating elliptical targets with square, rectangular and hexagonal point-nets*. Mineral Sci. Expt. Sta. Spec. Pub. 1–69, The Pennsylvania State University, University Park, Pa. 100 p.

Smith, A.Y. and R.J. Blackwell 1980. Development of an information data base for watershed monitoring. *Photogrammetric Engineering and Remote Sensing* 46(8), 1027–38.

Southard, R.B. no date. *Automated cartographic development in the United States of America*. Reston, Va.: U.S. Geological Survey.

Strahler, A.H. and J.E. Estes 1980. Incorporating collateral data in LANDSAT classification and modeling procedures. *Proc. 14th Internat. Symp. Remote Sensing of Environment*.

Teicholz, E. and J. Dorfman 1976. *Computer cartography world-wide technology and markets*. Newton, Mass.: International Technology Marketing.

Ternryd, C.O. 1972. A fundamental examination of the concept of accuracy in terrain data and earthworks calculations. *Proc. PTRC Seminar* 1–28.

Troeh, F.R. 1965. Landform equations fitted to contour maps. *Am. Jour. Sci.* 263, 616–27.

Turner, A.K. 1978. A decade of experience in computer-aided route selection. *Photogrammetric Engineering and Remote Sensing* 44(12), 1561–76.

Upton, W.B., Jr. 1955. *Index to a set of one hundred topographic maps illustrating specified physiographic features*. Reston, Va.: U.S. Geological Survey.

U.S. Geological Survey 1970. *Geological Survey manual*. App. to Part 800, Chap. 1, United States National Map Accuracy Standards, 9-2-70 (Release No. 1204).

U.S. Geological Survey 1980. *Geological Survey research 1979*. U.S. Geol. Survey Prof. Paper 1150, 447 p.

# 9

# GEOMORPHIC PROCESSES
# AND LAND USE PLANNING,
# SOUTH TEXAS BARRIER ISLANDS

## Christopher C. Mathewson and William F. Cole

### ABSTRACT

Barrier islands are active geomorphic areas that are also highly desirable development sites. Two basic classes of processes affect these areas: (a) hurricanes and storms, which are irregular, intermittent events; and (b) shoreline and aeolian processes, which are regular, daily processes. The intermittent events can drastically alter the island in a short time period, whereas the daily processes gradually modify the island. Land use planners and developers rarely consider the active processes or the impact of altering the geomorphology of the island. Studies conducted along separate areas of Padre Island, Texas, suggest that active geomorphic processes are different for each site. A study of the natural alteration of a part of the Padre Island Natural Seashore, using historical aerial photographs, shows that the island geomorphology follows a predictable cycle. The cycle is initiated when a hurricane breaches the foredune ridge. The resulting major foredune breach gradually heals as isolated dunes coalesce except at the downwind end where an aeolian chute forms. The chute funnels aeolian sand through the foredune ridge, forming an aeolian fan and back-island dune field. This aeolian transport system may exist for as long as 30 years. Eventually the chute closes and the island returns to pre-hurricane conditions. Land use decisions should contend with both irregular events and daily processes. It is impossible to predict where the next hurricane will strike, thus structures should either be designed to withstand the storm or as sacrifice structures. Once a hurricane breach forms, however, the location of the resulting aeolian fan can be predicted and the effect of aeolian transport can be incorporated into land use plans. Historical air photography of South Padre Island shows that hurricanes alter island geomorphology predictably and that daily processes do not effectively change this alteration. Hurricane hazard zones could easily be considered in land use plans to ensure responsible development of the area. To date, development on South Padre has disregarded the active coastal processes: residential zones have been built in washover channels, portions of the foredune ridge have been removed for hotel construction, construction siting has disregarded shoreline erosion, and aeolian sand is being removed for fill on the mainland. To maximize protection: from storm surge and scour, commercial structures should be founded on deep piles and located behind the foredune ridge; from washover channel erosion, channel sites should be maintained to relieve erosive pressure from the foredune ridge; from flooding, residential structures should be located on back-island areas away from washover channels. With low cost and little effort, geologic input can be incorporated into any plan and can make the barrier island a safer place to live.

## INTRODUCTION

The physical environment, unlike either the socioeconomic or biological environments, is often considered by man to be static and stable, whereas in fact this environment is highly dynamic. The land existing today along the Texas coast was produced through a continuous process of erosion and sedimentation. These processes are still active and capable of significantly changing the coast may times in a person's life span.

The Texas coast is characterized by flatlands crossed by numerous rivers, a scene very similar to one that has persisted over thousands of years. The typical coastline of Texas is one having sandy barrier islands facing the open Gulf of Mexico with protected lagoons between these islands and the mainland (Fig. 1). This barrier island/lagoon complex extends from Trinity Bay along the coast to near Brownsville. These beautiful islands are the basic physical environment upon which the recreational potential of the Texas coast rests.

Padre Island is presently undergoing extensive land and recreational development. Barrier islands, especially, have been the focus of rapid development because of their proximity to the ocean and appeal to sun and sea enthusiasts. Unfortunately for both

**Figure 1** Map of the Texas coastal zone showing the estuaries and lagoons that make up the environment.

seasonal tourists and permanent residents, the dynamic processes that have formed and continue to modify the barrier island are not generally considered by land use planners and builders. The result has been for many unwary consumers an economic and physical loss for which there is no adequate compensation (McGowen et al. 1970; Mathewson 1976).

Padre Island is the southernmost member of the chain of barrier islands that make up the Texas Gulf Coast shoreline (Fig. 2). To the north it is separated from Mustang Island by Corpus Christi Pass; on the south it is separated from the Rio Grande Delta Complex by Brazos Santiago Pass. Padre Island is 177 km long and is a nearly continuous network of beaches, dunes, and marshes interrupted only by Mansfield Pass. The island varies in width from 3·2 km in the north to slightly more than 0·8 km in its southern extremity. Vertical relief is predominately determined by the height of the larger dunes, which rarely exceed 12 m. Much of the island lies within 2·4 m of sea level. The distance from the mainland varies from about 8 km at the north end to about 16 km at the southern part.

Between Padre Island and the mainland lies Laguna Madre, a body of water that varies in width from 3·2 km at the north to a maximum of 12·8 km at the south and rarely attains a depth of more than 3 m. The Intracoastal Waterway, dredged by the Corps of Engineers along the entire length of the Texas Gulf Coast, traces a path within the lagoon parallel to the mainland. The Intracoastal Waterway is 60 m wide, and averages about 2·7 m in depth.

The climate of the south Texas coastal area is predominately semiarid subtropical. Annual temperatures average from 22 to 24°C; freezing is rare, occurring about once each year. Annual precipitation is extremely variable, ranging from 30 to 121 cm with an

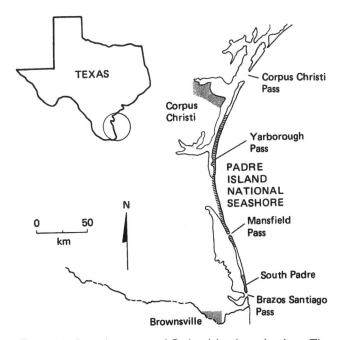

**Figure 2** Location map of Padre Island study sites. The National Seashore site is near Yarborough Pass and the other site is in the community of South Padre.

annual average of 66 cm. Precipitation is greatly biased by the number and intensity of tropical storms in a given year. Prevailing winds are from the southeast, variably punctuated with north to northeast winds of greater intensity usually associated with winter storms.

The first historical reference of man's uses of Padre Island dates to 1519, when Alonso Alvarez de Pineda set foot on the island and established Spain's claim. However, no attempt to colonize the island was made. The next contact the Spanish had with Padre Island was in 1553, when a fleet of Spanish galleons were sunk or ran aground during a storm. The Spaniards returned the next year and undertook a salvage effort that recovered about half the sunken treasure.

There seems to be some confusion in early historical accounts concerning the proper name of the island. In the south it was called Isla de Los Malquitos and Isla Blanca (either name being acceptable); in the north it was referred to as Isla de Corpus Christi.

The island was not referred to again by any of its names until the early 1800s, when the Spanish King, Charles IV, granted the island to Padre Nicolas Balli. The Padre gave the island both its name and its first industry – cattle raising. The island soon became known as "the Padre's Island." In 1827, a survey by order of Governor Fernandez was conducted to determine the extent of the land acquired by the Padre, who at the time of his death was a very wealthy and powerful man. More than a hundred families, in south Texas and across the Rio Grande into Mexico, trace their lineage back to the good Padre – a mark of his influence although not a credit to his profession.

In 1840, after the rebellion and the independence of Texas from Mexico, the southern part of the island was occupied by John Singer of the Singer sewing machine family of Boston. He grazed a herd of some 1500 head of cattle on the island but was forced to leave with the outbreak of the Civil War.

Following the vacancy caused by the Singer expulsion, Padre Island was deserted except for the U.S. Army. Between 1864 and 1865 some 5000 to 6000 troops were drilled and trained there. As isolated as it was, news traveled slowly and training continued for 33 days after Lee's surrender at Appomattox.

In the latter half of the nineteenth century, the cattle business was booming and in 1879 an Irish–American, Patrick Dunn, settled the island. The "Duke of Padre" as he was called soon owned most of the island and grazed several thousand head of cattle. In 1926 he sold his vast surface rights but retained the mineral rights, a holding still owned by members of the Dunn family.

From the days of Padre Balli to well after Dunn sold out, the cattle business ruled the island. With the increased demand for hydrocarbons in the 1940s and 1950s, however, petroleum companies began exploring and drilling in the area. When some of Laguna Madre and a small amount of Padre Island were affected by this drilling, public interest began to call for the creation of a park.

In 1962 Congress enacted legislation to obtain land on Padre Island and to establish Padre Island National Seashore. The National Seashore stretches some 105 km along the coast and provides camping, swimming, and fishing for visitors. The National Park Service is charged with the responsibility of (a) managing a geologically dynamic environment, and providing for (b) the development of potential energy deposits, (c) increased recreational uses, (d) the protection of the environment, and (e) wildlife habitat.

The island is presently open to most park visitors only at North and South Beaches and in the Malquite Beach day-use area because no roads have been constructed down the

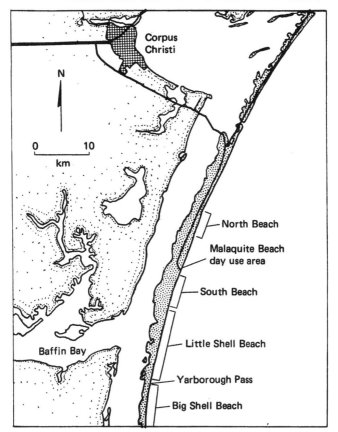

**Figure 3**  Padre Island National Seashore heavy-use areas.
Conventional vehicles and the more than 1,000,000 annual
visitors are confined to North and South Beaches and the
Malaquite Beach day-use area.

island (Fig. 3). Thus traffic must move on the beach. Little Shell and Big Shell Beaches contain shell pockets that are impassable in conventional vehicles. As a result, a majority of the park visitors are confined to about 24 km of the entire 105 km of National Seashore. Since park visitation now exceeds 1 million per year, this heavy use must be distributed throughout more of the available island space to protect the existing barrier island environment. A proposed master plan calls for the closure of beach access and construction of a shell road behind the foredunes as far south as Yarborough Pass.

In contrast to the National Seashore, South Padre Island, lying south of Mansfield Pass, has become a major coastal area development with beachfront hotels and condominiums, second houses, and other related recreational and community facilities. Development of South Padre Island as of August 1978 has disregarded the geological processes acting upon the island. Hotels and condominiums line the beach and, in many cases, the foredune ridge has been removed to make room for beachfront construction. Residential zones have been developed behind the hotel complexes. Many houses and resort cottages have little or no foredune protection; others are built in proven washover channels.

Although Padre Island is only one barrier island complex, two fundamental land use planning and public safety requirements exist. Along the Island, land use decisions and engineering construction must consider the diverse interactions between man and the geologic environment. The interrelationships between irregular, intermittent events (hurricanes and storms) and uniform, recurrent events (daily winds, tides, and tidal currents) must be understood to improve park management decisions and to provide adequate public safety along South Padre Island. The significant land use planning contrast between the National Seashore and South Padre Island is that one of these is intended for temporary camping (primitive) habitation at limited sites, whereas the other is designed for permanent (civilized) habitation throughout the area.

## ACTIVE GEOMORPHIC PROCESSES

In general, active geomorphic processes can be divided into two basic categories: (a) those related to hurricanes and other irregular, intermittent events and (b) those caused by daily winds and other regular, recurrent events. Hurricanes and storms are irregular in intensity and occur on an intermittent schedule, making them difficult to predict. Wind, on the other hand, is more regular in intensity and occurs more frequently, often having a cycle of a day or so.

Hurricanes have a significant impact on the geologic environment of Padre Island (Hayes 1967; McGowen et al. 1970). In general, a surge of high water washes through low places on the island, cuts passes, erodes the beach and dunes, and floods the back-island flats. These processes drastically alter the morphology of the island. The foredune ridge may be breached, new washover sands will cover parts of the vegetated flats, and washover channels may sever the island into a chain of islands. Once the storm passes, however, the uniform recurrent events once again become the significant geologic processes.

To identify the rate and magnitude of geomorphic changes occurring on Padre Island, historical geomorphic maps were constructed from aerial photographs. The island was divided into six geomorphic provinces: the beach, the foredune ridge, vegetated flats, wind-tidal flats, washover channels, and back-island dunes. Comparison of the maps reveals that daily shoreline processes alter island physiography rather slowly, but that large-scale, intermittent events produce rapid and drastic changes. These changes are described next.

### Padre Island National Seashore

A study of historical aerial photographs of the Yarborough Pass area suggests that uniform, recurrent events active on Padre Island are the significant processes to be considered in park management decisions and in the design and construction of engineering works (Mathewson et al. 1975; Mathewson 1977). In 1933 a hurricane formed both a major hurricane channel and a major foredune breach; another hurricane in 1936 probably reopened the breach in the foredune ridge. The 1937 aerial photogeologic map (Fig. 4) shows that the washover fan and the major hurricane breach have not revegetated. By 1943 the hurricane washover fan sediments were stabilized by grasses, the foredune ridge was discontinuous and partially vegetated, and the hurricane beach ridge had become an active dune field (Fig. 4). By 1953 the foredune ridge had reformed except for the southern end, where an aeolian chute had developed (Fig. 4). Between

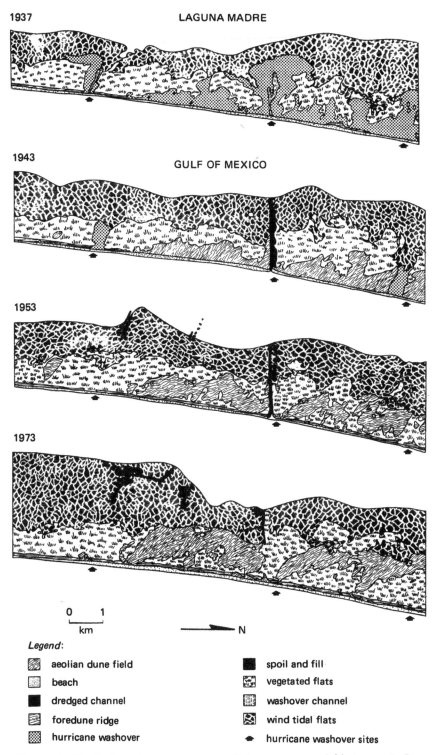

**Figure 4** Physical environment maps of the area around Yarborough Pass Padre Island National Seashore, showing the island environment in 1937, 1943, 1953, and 1973.

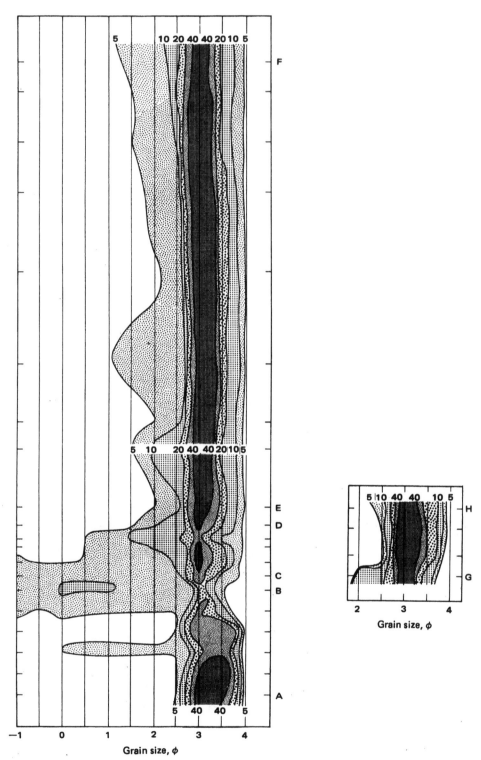

**Figure 5** Dowling diagram of the distribution of sediment size along a transect from the first bar offshore to the wind tidal flats. Note the influence of aeolian sorting. Figure 6 shows the map location.

138

1953 and 1973, sand was continually added to the aeolian fan through the chute, lengthening it in the downwind direction. By 1973 the chute had effectively closed and the aeolian fan had crossed the vegetated flats and encroached onto the wind tidal flats (Fig. 4). Vegetation is now slowly stabilizing the head of the aeolian fan. As the sand in the aeolian fan continues to move lagoonward, it apparently will migrate onto the wind tidal flats and eventually form the back island dunes. With a storm of the intensity of that in 1933, the chute may be reopened or the dune wall breached, initiating the cycle again.

Sediment data collected from the aeolian fan at Yarborough Pass further support the process described in the photogeologic analysis. Figure 5 is a Dowling (1977) diagram of the grain size distribution along a series of profiles starting at the first offshore bar,

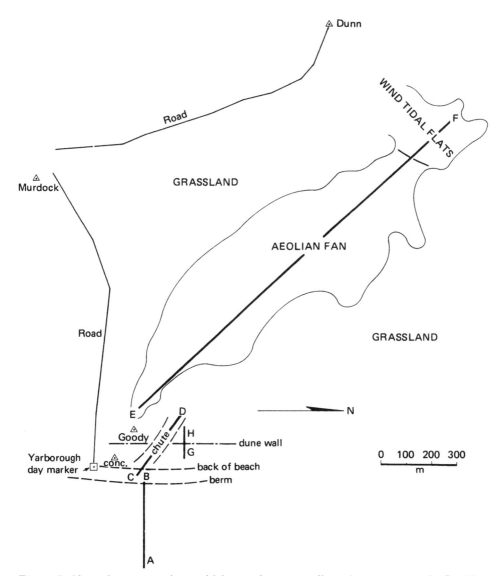

**Figure 6** Map of transects along which samples were collected to construct the Dowling diagram in Figure 5.

crossing the beach and extending through the aeolian chute to the wind tidal flats at the back of the island (Fig. 6).

These results show that the sand in the aeolian fan and wind tidal flats is usually distributed and transported by aeolian processes. Note the similarity in the characteristics of the sand in the foredune ridge (profile G–H) to that in the aeolian fan. The sharp shift

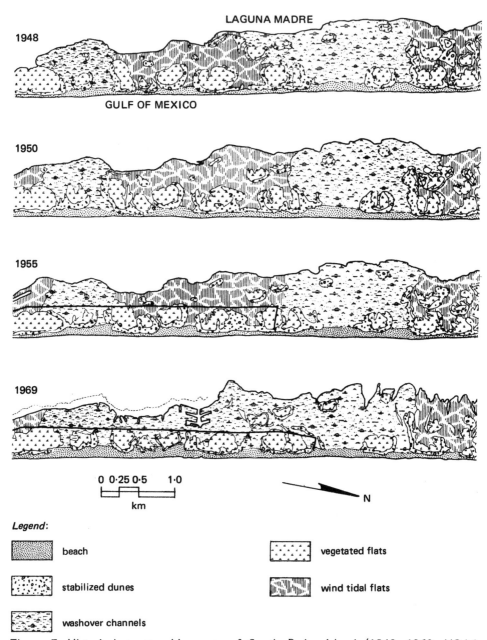

**Figure 7** Historical geomorphic maps of South Padre Island (1948, 1969–NOAA photography; 1950, 1955–USDA photography).

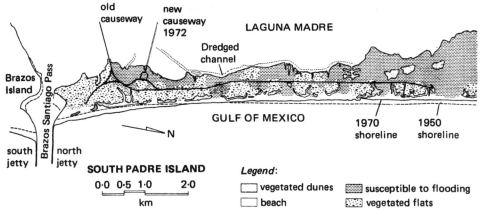

**Figure 8**   Shoreline changes along South Padre Island between 1950 and 1970.

to a coarser grain size in the head of the aeolian chute is indicative of higher wind shear caused by the funnel shape of the chute. This evidence is compatible with measured wind velocities at the head of the chute, which are 2 to 5 m/s higher than on the crest of the neighboring foredune. Mathewson (1977) suggested that the back-island dunes and aeolian fans on North Padre Island are the result of the same process: aeolian transport initiated by hurricane formation of a major foredune breach.

## South Padre Island

Studies of historical aerial photographs of a portion of South Padre Island suggest that irregular, intermittent events are the significant geomorphic process to be considered in land use planning. Figure 7 shows that the various geomorphic provinces exist in approximately the same position for the years 1948, 1950, and 1955, whereas the 1969 map is dominated by a large washover channel and many openings in the foredune ridge. The 1969 map reflects the amount of alteration that the island experienced during Hurricane Beulah, an intermittent event, which struck South Padre Island in the fall of 1967.

Geomorphic data suggest that active aeolian processes that would act to alter the island morphology are inhibited by a lack of a sediment supply. The foredune ridge is not reconstructed and large dune fields do not exist. The 1948, 1950, and 1955 aerial photographs show only minor accretion of the existing vegetated foredunes.

A comparison of 1950 aerial photographs with 1970 aerial photographs demonstrates that the significant daily process is shoreline erosion or accretion (Fig. 8). This geomorphic process, however, is limited to high-value beachfront developments and does not represent a general risk to the entire island as does the hurricane.

## LAND USE PLANNING

It is interesting to note that the analysis of historical aerial photographs near Yarborough Pass, Padre Island National Seashore, and on South Padre Island suggest that these two sites are significantly different. Geomorphic changes in the National Seashore are due predominately to aeolian transport of sand, whereas those at South Padre are due predominately to hurricane and storm waves. As a result, land use planning decisions must not only consider the intended land uses but also the various geologic processes.

## Padre Island National Seashore

The rapid reconstruction of the continuous foredune ridge, except at the aeolian chute, makes it impossible to predict where the next hurricane will breach the dunes. It is therefore not possible to establish high-hurricane-risk zones for the back island areas since no obvious washover site is preserved. The hurricane must be considered as a "rare" event that provides limited design alternatives for any park improvements. Structures should be designed and built to withstand scour, flooding, wave attack, and high winds or be sacrificed to the storm. Management choices should be based on economic considerations, cost-risk analyses, potential environmental hazards, and the goals of the park plan.

Aeolian processes, however, are preserved and can be evaluated through the use of historical aerial photographs. Once a hurricane has formed a major foredune breach, it is possible to predict where back-island areas will be affected by aeolian transport. The development of dune fields in the area progress through five stages (Fig. 9). The initial morphologic features include the beach, the foredune ridge, and vegetated flats which grade lagoonward into wind tidal flats. Sediment is deposited on the beach by wave action, sorted and transported by the prevailing southeast winds, and finally, deflected northward by the foredune ridge.

Aeolian transport across the island is inhibited when the foredune ridge and back-island flats are stabilized by vegetation. However, hurricane attack of the foredune ridge strips away the stabilizing vegetation and forms a hurricane beach, allowing aeolian processes to become highly effective (Fig. 9, stage 2). In the case of major and minor hurricane channels, the breach is narrow and such channels heal quickly when compared to a major foredune breach. In the area of a major breach, hurricane waves form a hurricane beach.

The hurricane beach is topographically higher than the surrounding flats and above the capillary fringe of the island freshwater lens. This higher elevation allows the sand to drain; thus, revegetation is slower and aeolian processes become the major transport mechanism. Reconstruction of the foredune ridge by aeolian processes, as described by Bagnold (1941) and Fisk (1959), closes the major foredune breach by forming scattered dunes which eventually coalesce. At the south end of the breach, high sediment transport and high wind shear apparently keep sand from being deposited (Fig. 9, stage 3), and an aerodynamically stable chute forms. The aeolian chute continues to serve as a conduit for sand that is transported to the back-island areas, producing the aeolian fan (Fig. 9, stage 4). At Yarborough Pass, the chute is aligned with the prevailing wind and remained a transport route for windblown sand for more than 30 years.

Eventually the chute becomes inefficient because of encroachment of the vegetated foredune ridge and closes (Fig. 9, stage 5). Following the closing of the chute, the source of sediment for the aeolian fan is lost and the sand migrates across the islands as a dune field and eventually becomes the back-island dunes. This study suggests that the large active dune fields in the National Seashore were formed through long-term aeolian processes rather than through hurricane processes alone, as proposed by Boker (1953), Blankenship (1953), and Hayes (1967).

Engineering works and park management decisions on Padre Island must contend with both categories of active processes. With respect to hurricanes, management

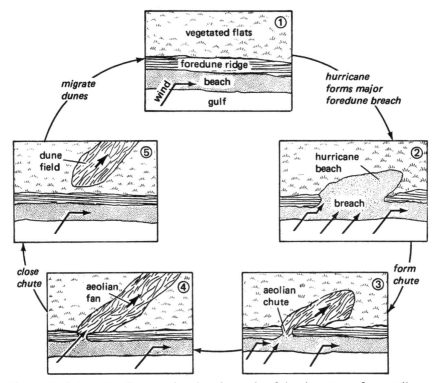

**Figure 9** Schematic diagram showing the cycle of development of an aeolian fan and fields, based on studies of historical aerial photographs near Yarborough Pass, Padre Island National Seashore. Initially, a hurricane breaches the foredune ridge and forms a hurricane beach. The breach gradually heals until the chute forms. Continuous sand transport through the chute forms the aeolian fan. Once the chute closes, the dune field migrates downwind.

decisions are usually limited to evacuation of personnel and visitors and to the closing down of operations. In terms of the engineering impact of hurricanes, structures can either be designed and built to withstand the storm or as sacrifice structures, depending upon economic and environmental considerations. Roads actually fit both categories, in that they are washed out in areas of high erosion and simply drowned by flooding in other areas. The net result of a hurricane is the cost of repair once the storm has passed, costs that should be incorporated into the design of the structure.

Rebuilding and repair after major storms should consider the post-storm island morphology with respect to the potential of modification by aeolian processes. It is impossible to predict the site of the next major hurricane breach in the foredune ridge; however, once the breach occurs the location of the related aeolian chute and fan can be predicted if it should develop. Thus post-storm reconstruction should consider the effect of aeolian transport as an economic parameter in the management decision.

## South Padre Island

South Padre receives the full fury of the storm surge during the passage of hurricanes. Using the classification of the *Model Minimum Hurricane-Resistant Building Standards*

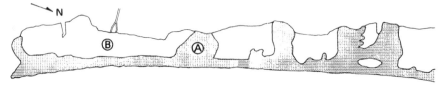

**Figure 10** Hurricane hazard zones on South Padre. Zone A is susceptible to scour, battering with debris, flooding, and high winds. Zone B is protected from scour by the foredune but is susceptible to battering, flooding, and high winds.

*for the Texas Gulf Coast* (Texas Coastal and Marine Council 1976), the barrier island can be divided into two hurricane zones (Fig. 10). Zone A consists of areas open to the full impact of the storm surge, which includes the beach, the foredune ridge, and back-island areas directly exposed to the sea by washover channels. Structures built in zone A are subject to high-velocity winds, battering with debris, flooding, and scour. Areas protected by a continuous foredune ridge are not open to the high-velocity flow that causes scour, but they are still subjected to high winds, battering with debris, and flooding. Zone B would include back-island areas protected by a foredune ridge and not directly exposed to washover channels.

Low elevations combined with bulldozed openings in the foredune ridge on South Padre Island greatly increase the probability of back-island flooding and scour. The maximum elevation of the foredune ridge along South Padre Island is only 2·1 m above mean sea level, whereas the maximum storm surge height of hurricanes Carla and Beulah exceeded 4 m. The practice of bulldozing down the foredune ridge for hotel construction creates potential washover channels for future storms. This endangers back-island areas normally protected by a foredune ridge. In August 1978 tropical storm Amelia backed up sewer lines and caused flooding even though this was only a moderate storm. Although the entire island may be inundated when a hurricane strikes, the destructive force of the storm surge may be diminished by the establishment of a stable, continuous foredune ridge and vegetative cover on dunes and flats.

Although not as catastrophic as hurricanes, daily processes also have a lasting impact on the coastal regime. Shoreline changes result from the interaction of tides, sediment supply, storms, and sea-level fluctuations. The predominant southeasterly winds along the South Texas coast control the direction of the longshore currents as well as the angle at which waves attack the beach. LANDSAT photography indicates that the southern portion of the Texas coastline receives its sediment from longshore currents flowing northward from the Rio Grande delta.

Shoreline erosion cannot be prevented by man-made structures if there is not enough sediment in the system to allow deposition. To alleviate the problems of shoreline erosion and a low foredune ridge, the sediment supply should be renourished. Sand should be placed along the beach and allowed to be reworked by waves, and sand should also be placed in low spots in the foredune ridge and then stabilized by vegetation. There are several sources for sand on South Padre Island:

(a) Presently, maintenance crews are removing sand dunes that are blowing across Park Road 100 north of the town and using the material as fill in construction sites on the mainland. This sand should be redistributed along the receding shoreline.

(b) Sand accumulating on the south side of the Brazos Santiago Pass or Mansfield Pass jetties could be used to renourish the beach.

In addition to renourishing the sediment supply, other methods of stabilizing the island and preventing back-island erosion and scour during hurricanes include prohibiting building and traffic along the beach, preventing foredune destruction for construction purposes, and establishing a dense vegetative cover where possible.

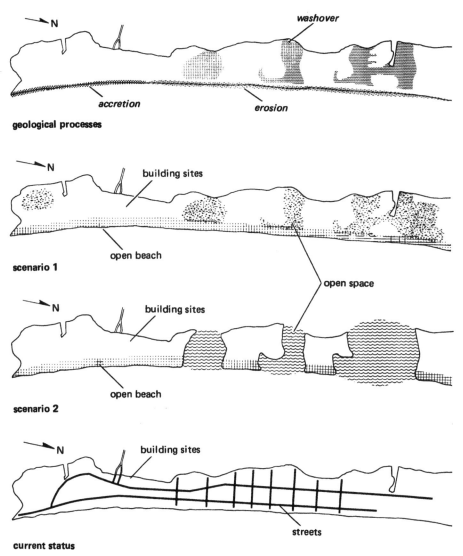

**Figure 11**  South Padre land use planning scenarios based on the geologic processes. The beach and washover areas are maintained as open spaces in order to minimize losses due to storms. The open washover channels act to relieve storm surge pressures and thereby help protect the foredunes. Note that the current status of land use on South Padre does not recognize the geologic processes.

Based on an understanding of the geological processes affecting South Padre Island, two planning scenarios are suggested that would greatly enhance public safety (Fig. 11). Scenario 1 identifies the hazards of shoreline erosion, destruction of structures in washover channels, inundation of the island in the event of a severe hurricane, sewer backup and water shortages, and temporary halt of vehicular movement. Under this scenario washover areas are designed as open spaces and residential development is confined to the stable portions of the island (i.e., back-island areas protected by a foredune ridge). It is recommended that performance standards for building construction and programs for renourishment of the depleted sediment supply be adopted by the city. Scenario 2 includes the conditions of scenario 1 but also considers potential separation of the island areas between washover channels, which may occur in the event of a particularly disastrous hurricane. Emergency facilities, residential zones, and high-cost commercial buildings are confined to the most stable portions of the island. A comparison of the current status and the geologic processes maps reveals that geology has not been considered in the development and land use planning on South Padre Island.

## CONCLUSIONS

The application of geologic and geomorphic process information into land use decision making involves more than simply identifying hazardous areas. Any scenario that proposes restrictive land use, such as open space in washover areas or downwind of aeolian fans, must consider the constitutional rights of private land ownership. The site best suited to the direct application of geologic design is a site under single ownership, such as the case in the National Seashore. In South Padre, where multiple ownership exists, the concept of a "joint land corporation" or a "Padre Island Development District" in which landowners' shares are based on the value of their land could allow the same design flexibility afforded a single owner.

In addition to the problem of land ownership is the problem of defining land use zones rather than total density zones. It is impossible to recover any investment if density decreases caused by establishing open spaces are not compensated by allowing higher-density uses in other areas. Classical approaches to urban planning must also respond to the availability of geologic information.

## REFERENCES

Bagnold, R.A. 1941. *The physics of blown sand and desert dunes.* New York: William Morrow, 265 p.

Blankenship, W.R. 1953. Sedimentology of the outer Texas coast. Master's thesis, Univ. Texas, 72 p.

Boker, T.A. 1953. Sand dunes on northern Padre Island. Master's thesis, Univ. Kansas.

Dowling, J.J. 1977. A grain size spectra map. *Jour. Sed. Petrologists* 47(1), 281–84.

Fisk, H.N. 1959. Padre Island and the Laguna Madre flats, coastal south Texas. 2nd Coastal Geol. Conf., La. State Univ., Baton Rouge, 103–51.

Hayes, M.O. 1967. *Hurricanes as geological agents: case studies of hurricanes Carla, 1961, and Cindy, 1963.* Univ. Texas, Bur. Econ. Geology Rept. Invest.

Mathewson, C.C. 1976. Active geologic processes on Padre Island National Seashore – impact on park management. Abs., First Conf. Sci. Research in the National Parks, New Orleans, La., Nov.

Mathewson, C.C. 1977. Beach washouts, an adverse effect of artificial dune construction. *Assoc. Eng. Geologists Bull.* **14**(1).

Mathewson, C.C., J.H. Clary and J.E. Stinson II 1975. Dynamic physical processes on a south Texas barrier island – impact on maintenance. Ocean 1975, Conf. of Engineering in the Ocean Environment and 11th Ann. Mtg. of the Marine Technol. Soc., San Diego, Calif., Sept. IEEE Pub. 75 CHO 995-1 OEC, 327-30.

McGowen, J.H., C.G. Groat, L.F. Brown, Jr., W.L. Fisher and A.J. Scott 1970. *Effects of hurricane Celia – a focus on environmental geological problems of the Texas coastal zone.* Univ. Texas at Austin, Bur. Econ. Geology, Geol. Circ. 8-3, 35 p.

Texas Coastal and Marine Council 1976. *Model minimum hurricane-resistant building standards for the Texas Gulf Coast.* General Land Office of Texas, Austin, Tex.

# 10

# MAN-MADE STRUCTURES AND GEOMORPHIC CHANGES SINCE 1876 ALONG THE OHIO SHORE OF LAKE ERIE

## C.H. Carter, D.J. Benson, and D.E. Guy, Jr.

### ABSTRACT

The length of the 300-km-long mainland shore which receded at less than 0·3 m/yr increased from 151 km in the early period to 171 km in the late period. In addition, the length of shore fronted by wide ($\geq$ 15 m) beaches decreased from 64 km in 1876–77 to 35 km in 1968, and the length of shore without a beach increased from 84 km in 1876–77 to 112 km in 1968.

These changes have been caused largely by man-made structures. Harbor structures – mostly jetties – have caused accelerated recession, whereas shore-protection structures – mostly groins and seawalls – which have increased in number from about 60 in 1876–77 to about 3600 in 1973, have caused decelerated recession. The effect of the shore-protection structures has more than offset the effect of the harbor structures. Seawalls, by directly armoring the shore from waves, and groins, by trapping sand and thus reducing the wave energy reaching the shore, have also led to many of the changes in the beaches by reducing the volume of sand supplied by shore erosion. Thus even though the best natural form of shore protection – a beach – is diminishing, the shore is receding less rapidly because of the shore-protection structures.

## INTRODUCTION

Retreat of the last Wisconsinan glacier about 12,600 years B.P. allowed discharge of the last Lake Erie Basin Pleistocene lake northeastward across the Niagara Escarpment. Because this outlet was about 40 m below present lake level, the level of the lake was lowered by this amount, forming early Lake Erie. Isostatic rebound of the escarpment then caused a rise in lake level to the present elevation of about 174 m above sea level. Man-made development of the 300-km-long mainland lake shore (Fig. 1) began in the late 1700s and early 1800s in the port towns, but by the early 1900s a large portion of the coastal zone had been settled. Lakeshore processes were soon apparent to the coastal inhabitants (Whittlesey 1838, p. 53):

> When the first settlers of the Western Reserve came along the Ohio shore, in 1796, the sandy beach of the lake was occupied as a road throughout, and was used for that purpose east of Cleveland many years.
>
> At the present time, the encroachment upon surveyed lots, between the Cuyahoga and Chagrin rivers, is from 10 to 20 rods [50 to 100 m].

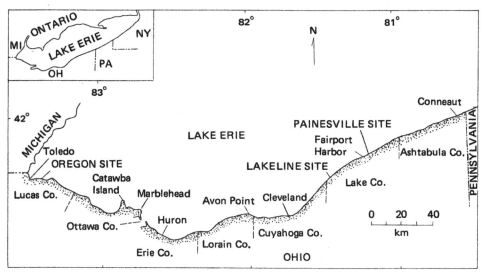

**Figure 1** The Ohio mainland shore of Lake Erie.

If we except a short distance along the shore west of Conneaut harbor, about 5 miles [8 km] next westerly of Fairport, and 20 miles [32 km] rock coast between Cuyahoga and Black Rivers, the entire shore from the State line to the lime rock near Huron, has lost an average of 8 rods [40 m] in width. The immediate bank is composed of loose earthy materials incapable of resisting the action of the waves, with the above exceptions.

In the early 1970s – a time of abnormally high lake levels – the Ohio Division of Geological Survey began detailed studies of each of the eight counties that border the lake in order to document and interpret the historical changes and thus better understand the shore and nearshore processes and predict future changes. For these studies, field

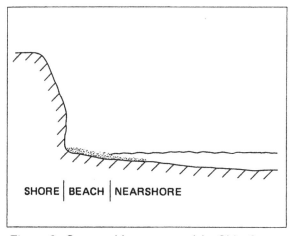

**Figure 2** Geomorphic zones used in Ohio Survey studies.

data were collected from the nearshore, beach, and shore zones (Fig. 2), and U.S. Lake Survey field sheets from 1876–77 and aerial photographs from the late 1930s and 1973 were used to map recession lines – commonly the intersection between the nearly horizontal upland surface and the sloping surface that fronts the lake – as well as changes in beaches and man-made structures. These data were then combined and interpretations were made for the changes.

The most significant aspect of our work in an applied sense has been the tremendous effect of man-made structures – harbor structures and shore-protection structures – on the shore. In a way, the effect of the structures has mimicked the Holocene geologic changes. For example, along the prehistoric shore the most resistant stretches form the headlands such as those at Avon Point and Marblehead, and the least resistant stretches form the embayments such as those at Cleveland and Huron. The man-made structures, by changing the nature of the shore or by changing the flux of wave energy that reaches the shore, have caused headlands and embayments to form and in addition have altered other processes.

## General procedure

Field studies were made in the 1970s. The shore deposits were mapped by boat with detailed sections measured from place to place to maintain continuity. Man-made structures were inventoried on foot and in boats. Beach observations were made and samples were taken at the time the nearshore profiles were run with continuously recording echo sounders. SCUBA divers mapped the bottom along the profile lines.

Map and photographic work closely followed the field work. U.S. Lake Survey maps from 1876–77, U.S. Agricultural Adjustment Administration photographs from 1937–40, and Ohio Department of Transportation photographs from 1973 were used to map the beaches, man-made structures, and the recession lines. Preparation of the recession-line map was essentially a three-stage process: (1) preparation from the 1973 photographs of an overlay on acetate which shows the position of the 1973 recession line as well as certain roads and prominent geographic features that can be used to ensure accurate correlation and alignment of the overlapping aerial photographs; (2) projection and enlargement of the 1876–77 maps to the same scale as the 1937–40 photographs and transfer of the 1876–77 recession line to the 1937–40 aerial photographs; and (3) projection and enlargement of the 1937–40 aerial photographs (with the 1876–77 recession line) to the same scale as the 1973 aerial photographs and transfer of the 1876–77 and 1937–40 recession lines to the 1973 acetate base map.

The recession rate at a given location can be determined from the recession-line map by measuring the perpendicular distance between recession lines for two different years and dividing this distance by the period defined by the recession lines. In measuring recession rates along the shore a scale that showed the map distance for the specific recession-rate classes within a given time period was used. In this way the scale could be moved between two recession lines, and the breaks between the recession rate classes could be marked off for later measurement of the distance along the shore.

Accurate horizontal control among the maps and photographs during projection and enlargement was maintained by correlation of geographic control points. With increased shore development there are a much greater number of control points on the 1937–40 photographs which can be accurately matched with the 1973 aerial photographs. Once control points on the map and photo or photo and photo were matched, many other natural and cultural features could be used to give more precise correlation.

Accuracy of position of the recession lines is unknown. However, there is much indirect evidence that suggests that the lines are accurate to within several meters at a given location. Moreover, the greatest source of possible error appears to be human; if this is so, errors in drawing the recession lines will probably be random, so that deviations will tend to cancel out along the shoreline. Overall, the map and photograph scales were quite consistent, and the transfer of the recession line from the 1876-77 maps to the 1937-40 photographs and from the 1937-40 photographs to the 1973 photographs was done at essentially the same amount of enlargement.

Implicit in our work with 35- and 60-year periods is the understanding that the changes that we have been able to document are for the most part markedly nonlinear and, in the case of the man-made structures and beaches, incomplete. For example, a period of high lake levels is usually accompanied by a marked increase in the number of shore-protection structures and by increased recession rates. Also, because of the length of the periods, we are unable to fully document changes in the man-made structures and the beaches.

## Previous work

This paper is an outgrowth of the Ohio Survey's county shore erosion reports (e.g., Carter 1976b; Benson 1978; Carter & Guy 1980). Detailed information concerning the methods, findings, and other data can be found in these reports.

## PHYSICAL SETTING

### Meteorology and limnology

The climate is markedly influenced by Lake Erie. The following information from the Cleveland summary of the Environmental Data and Information Service (1978) applies in general to the Ohio coastal zone:

> Cleveland's climate is continental in character but with strong modifying influences by Lake Erie. West to northerly winds blowing off Lake Erie tend to lower daily high temperatures in summer and raise temperatures in winter.... Summers are moderately warm and humid with occasional days when temperatures exceed 90°F [32°C]; winters are reasonably cold and cloudy with an average of five days with sub-zero temperatures. Weather changes occur every few days from the passing of cold or warm fronts and their associated centers of high and low pressures.... As is characteristic of continental climates, precipitation varies widely from year to year; however, it is normally abundant and well distributed throughout the year with spring being the wettest season. Mean annual precipitation is about 34 inches [86 centimeters].

The prevailing winds are from the southwest. Because of the orientation of the lake, the most effective winds come most frequently from the westerly quadrants, with a secondary maximum from the northeast (U.S. Weather Bureau 1959, p. 13). The strongest winds usually are generated by low-pressure systems.

These LOWS originally form in three areas: (1) Texas and New Mexico; (2) the Central Rocky Mountains and Great Plains; and (3) the Pacific Southwest. The movement of storms from all three regions is similar, from the Middle West to the Great Lakes. The season for storms from these regions is generally from October through May (U.S. Weather Bureau 1959, p. 4).

Of particular significance to the Ohio shore of Lake Erie are the lows that pass to the south of the lake. These lows generate northeast winds which, because of the long fetch, produce the largest waves along the western half of the lake.

Surface waves are the most important coastal process along the Ohio shore. Wave hindcast data for Monroe, Michigan, and Cleveland provide an estimate of wave height and direction along the western and eastern portions of the shore (Table 1). Storm waves from the northeast commonly have periods of 5 to 7 s and heights in the breaker zone of about 1 m. In general, east of Avon Point waves generated by the prevailing westerlies combined with the northeast-trending shore account for the net west-to-east transport of sand; west of Avon Point waves generated by the less frequent northeasters, because of greater fetch, account for the net east-to-west transport of sand.

Natural lake-level changes (Fig. 3) can be divided into three types: short term (changes within a few days or less), medium term (changes within a year), and long term (changes over a few years or more). The short-term changes due mostly to wind-stress buildup, barometric pressure changes, and seiche activity – astronomic tides are generally no more than several centimeters in amplitude – are most pronounced at the confined ends of the lake near Buffalo and Toledo. An intense northeaster can raise the lake level 2 m at Toledo and 1 m at Marblehead. Medium-term (seasonal) changes are lakewide in effect and are caused primarily by differences in rates of runoff, evaporation, and evapotranspiration. A typical seasonal cycle for Lake Erie shows a high in June–July and a low in January–February; the mean difference in elevation between the high- and low-water stages is about 0·4 m (Lake Survey Center 1973). Long-term (historical) changes are caused largely by major weather changes; the changes that affect precipitation and evaporation are most important. The variations in lake level can be quite pronounced;

**Table 1**   Wave height data for ice-free period, April 1 to December 1 (from Saville 1953, p. A11, B15)

| City | Height (m) | NE | ENE | E | ESE | SE | SSE |
|---|---|---|---|---|---|---|---|
| | | | | **Direction of wave (average % of time)** | | | |
| Monroe[a] | 0·15–0·9 | 2 | 2 | 2 | 3 | 2 | < 1 |
| | 0·9–1·8 | < 1 | < 1 | < 1 | < 1 | < 1 | – |
| | > 1·8 | – | < 1 | < 1 | – | – | – |
| | | W | WNW | NW | NNW | N | NNE | NE |
| Cleveland[b] | 0·15–0·9 | 1 | 2 | 2 | 3 | 3 | 3 | 2 |
| | 0·9–1·8 | < 1 | 1 | 1 | 1 | < 1 | 1 | < 1 |
| | > 1·8 | < 1 | < 1 | < 1 | < 1 | – | – | < 1 |

[a] Lake calm or wave height less than 0·15 m 87% of the time.
[b] Lake calm or wave height less than 0·15 m 79% of the time.

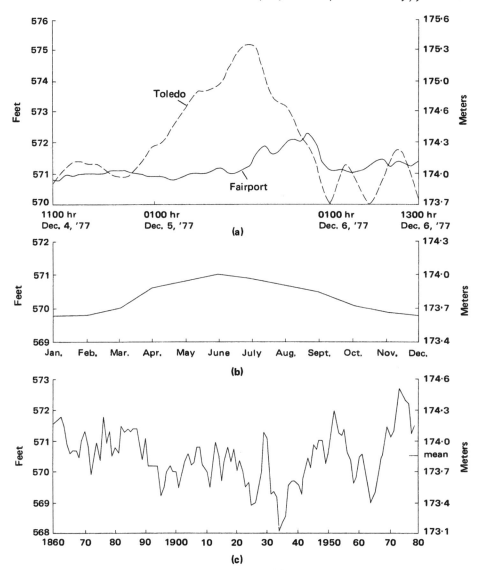

**Figure 3** Lake level fluctuations on Lake Erie: (a) wind setup, (b) annual cycle, and (c) annual means.

for example, in December 1934 the mean lake level was about 173·0 m, whereas in December 1972, after several long-term up-and-down cycles, the mean lake level was about 174·5 m (Lake Survey Center 1973).

## Geography and geology

The mainland shore from Toledo to Huron (Fig. 4) consists of low-relief (< 2 m) barrier beaches and laminated clay banks except for the Catawba Island–Marblehead area, which is made up largely of moderate-relief (3 to 6 m) dolostone and limestone slopes and bluffs. The shore from Huron to Conneaut consists of moderate-to-high relief (3 to 20 m)

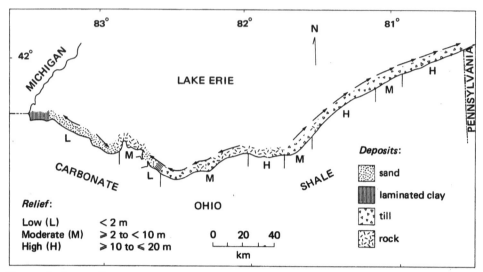

**Figure 4** Generalized map of shore deposits in the wave erosion zone, relief, net long-shore transport directions (arrows), and bedrock geology along the Ohio mainland shore.

shale and/or till slopes and bluffs commonly capped by stratified clay, silt, and/or sand. Shale makes up most of the shore between Avon Point and Cleveland. Excluding man-made structures and fill, about 47% of the length of the shore in the wave erosion zone is till, 26% is rock, 22% is sand, and 5% is laminated clay.

Beaches occur as a fragmented band that fronts about 45% of the mainland shore. The beaches are commonly narrow (< 15 m wide) and consist primarily of sand, although there are cobble pocket beaches in places where the shore is composed of rock. Adjacent to the beaches are the man-made structures, which consist largely of groins and seawalls. In general, the longest continuous beaches are found where there are the fewest groins and jetties.

Nearshore slopes are gentle; the slopes within 600 m of the shoreline are usually less than 1°. The bottom is generally made up of sand and gravel less than 2 m thick near the shore and of rock, till, glaciolacustrine clay, or silt farther offshore. Offshore bars are commonly associated with the largest beaches.

## Processes

There are two principal shore erosion processes: wave erosion and mass wasting. Of the two, wave erosion is by far the more significant because the amount of mass wasting is largely dependent upon the amount of wave erosion. Wave erosion of in situ or slumped material from the base or toe of the bank, bluff, or slope reduces the resisting moment and the shore becomes more susceptible to downslope movement. Thus the shore does not reach a long-term equilibrium; rather, it is in an unstable state. On the other hand, if there were no wave erosion and therefore no significant erosion process to remove material from the base of the slope, the slope would reach a stable angle of repose and would be affected only by relatively slow processes such as creep. Thus the shore would reach a long-term equilibrium; it would be in an essentially stable state.

## Recession rates

Recession rates appear to follow the expressions derived by Sunamura (1977, p. 613):

$$\frac{dx}{dt} \propto F \qquad \ln\left(\frac{f_w}{f_r}\right) \propto F$$

where $dx/dt$ = erosion rate
$F$ = wave erosion force
$f_w$ = wave-related erosion forces
$f_r$ = resisting force of the shore deposits

The relative resistance of the shore deposits appears to control the recession rates along the Ohio shore. For example, in the natural cohesive deposits, recession rates increase from rock (dolostone-limestone, shale) to till to laminated clay (Carter 1976a). Shore-protection structures – mostly seawalls – by acting as a yet more resistant shore deposit further reduce recession rates by armoring the shore behind them.

Numerous factors have a bearing on the wave-related erosion forces reaching the shore. Among the most important are exposure (fetch and orientation of the shore), nearshore slope, and beach width. However, for a given wave climate and physical setting, the two most important factors are ice and lake level. In general, nearshore ice limits wave erosion by damping waves and preventing them from reaching the shore or by armoring the shore. Lake level affects wave erosion by influencing the distance from shore at which the waves break. For a given wave, the higher the lake level, the closer the wave breaks to the shore and thus the greater the amount of wave energy reaching the shore. The most intense wind storms and thus largest waves and greatest wave energy on Lake Erie occur by coincidence when the lake level is lowest (late fall to early spring) and when there is ice on the lake (about mid-December to mid-March) (Fig. 5). The shore-protection structures – mostly groins – by trapping sand and thus building a wider beach further reduce recession rates; however, they can at the same time increase recession rates by causing a downdrift decrease in beach width.

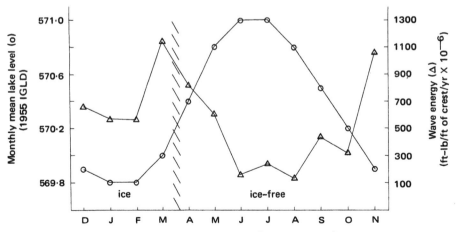

**Figure 5** Monthly wave energy at Cleveland (Saville, 1953), monthly mean lake level, and annual period of nearshore ice.

With respect to the resistance of the shore and wave-related erosion, the resistance of the shore appears to be the most crucial variable. For example, the rock headlands at Avon Point and Marblehead are exposed to the greatest wave-related forces along the entire shore, yet recession rates are usually less than 0·3 m/yr, whereas the laminated clay banks at Maumee Bay are exposed to the smallest wave-related forces along the entire shore, yet recession rates are commonly 2 to 3 m/yr. Obviously, for a given shore deposit and physical setting, the greater the wave-related forces, the greater the recession.

## HISTORICAL SHORE AND BEACH CHANGES

### Man-made structures

There are two principal types of man-made structures along the shore: harbor structures and shore-protection structures. The harbor structures, which consist largely of jetties and breakwaters, were built to keep the river mouths open and to reduce the effect of waves and swell. The shore-protection structures, which consist largely of groins and seawalls, were built to reduce shore erosion. The seawalls, by armoring the shore, increase the resistance of the shore to waves, whereas the groins, by trapping sand, decrease the amount of wave energy reaching the shore.

The first harbor structures were jetties built in the mid-1820s. Most of the jetties trapped sand transported in the nearshore zone so that their lengths were increased from time to time to keep the harbor entrances open. For example, one of the biggest harbor structures along the Ohio shore was begun in 1825 at Fairport Harbor. By 1893 the west jetty was about 720 m long (U.S. Army, Chief of Engineers 1932, p. 10). A land-based breakwater was then added to the west of the jetty; in 1949 this structure was about 1180 m long (U.S. Army, Corps of Engineers 1950, plate 7), making the effective length of the structures perpendicular to the shore about 1500 m. These structures trapped sand

Table 2   Large structures along the Ohio shore (from Hartley 1964)

| Location | Type | Length normal to shore (m) |
|---|---|---|
| Port Clinton | Jetties | 380 |
| Cedar Point | Jetty | 1830 |
| Huron | Jetties | 910 |
| Vermilion | Jetties | 305 |
| Beaver Creek | Jetty | 240 |
| Lorain | Breakwaters | 1550 |
| Cleveland[a] | Breakwater | 1100 |
| White City Park | Breakwater | 460 |
| Eastlake | Jetties | 340 |
| Mentor Harbor | Jetties | 150 |
| Fairport | Breakwaters | 1310 |
| Ashtabula | Breakwaters | 1680 |
| Conneaut | Breakwaters | 1420 |

[a] Structure lies parallel to the shore.

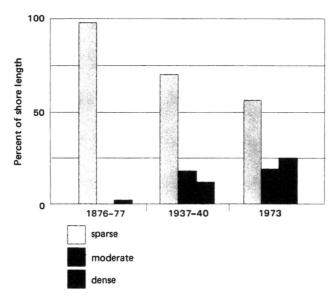

**Figure 6** Changes in structure density along the Ohio mainland shore. *sparse*, $< 1/3$ of shore protected by structures; *moderate*, $1/3$ to $2/3$ of shore protected by seawall-like structures or by closely spaced groins with trapped sand beaches; *dense*, $> 2/3$ of shore protected by seawall-like structures or by closely spaced groins with trapped sand beaches.

transported from the west at a rate of about 95 000 m³/yr between 1911 and 1945 and at a rate of about 67 000 m³/yr between 1945 and 1958 (Bajorunas 1961, p. 333, 334). Hartley (1964) lists thirteen large structures ("which have had a measurable effect on half a mile or more of shore") along the Ohio shore of Lake Erie; eleven of these structures are harbor structures (Table 2).

Development of the shore has been accompanied by an associated increase in the number of shore-protection structures and in the density of shore protection. For example, in 1876–77 there were about 60 shore-protection structures, in 1937–40 about 1400, and in 1973 about 3600. Associated with the increase in the number of structures is an increase in structure density (Fig. 6). In contrast to the harbor structures, these structures are small – generally no more than 30 m long – and have been built wherever shore protection was wanted.

## Beaches

The distribution and size of beaches has clearly changed along the shore from 1876–77 to 1968 (1968 was chosen for comparison because of a similar lake level). In 1876–77 lake levels were between 174·1 and 174·5 m and there were 61 stretches without beaches, whereas in 1968 the lake level was about 174·1 m and there were 340 stretches without beaches. Associated with this increase in the number of stretches without beaches has been a decrease in beach width, as evidenced by the decrease in the length of shore fronted by wide ($\geq 15$ m) beaches (Fig. 7). Even in the years of lower lake level such as

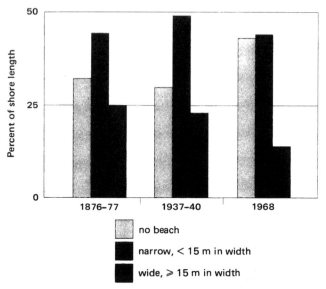

**Figure 7**    Changes in beach widths along the Ohio mainland shore.

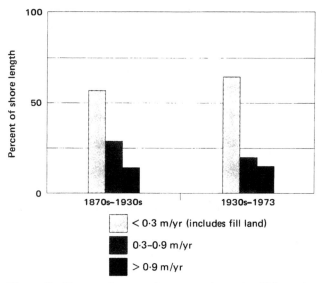

**Figure 8**    Changes in recession rates along the Ohio mainland shore.

1937–40 (lake level during the photographic flights ranged from about 173·4 m to 173·9 m) the length of shore fronted by wide beaches was less than in 1876–77.

## Recession rates

Recession rates in the two long-term periods also have changed in a less pronounced but still well defined way (Fig. 8). For example, 57% of the shore receded at less than 0·3 m/yr in the early (1876–77 to 1937–40) period, whereas 64% of the shore receded at this rate in the late (1937–40 to 1973) period. On the other hand, 14% of the shore receded at greater than 0·9 m/yr in the early period and 15% in the late period. Moreover, the changes in recession rates have commonly not been uniform along the shore; this lack of uniformity has led to a more irregular shape (Figs. 9 to 12). The most irregular stretches are characterized by nearly complete seawall protection with unprotected gaps, as at Oregon, or by sparsely protected stretches characterized by a few 20- to 30-m long groins, as at Painesville-on-the-Lake, or by a harbor jetty. However, where the shore is unprotected or where the shore is uniformly protected the shore has retreated fairly uniformly (Fig. 13).

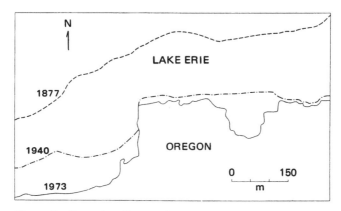

**Figure 9** Recession lines at Oregon, Ohio.

**Figure 10** Vertical aerial photograph taken in 1973 at Oregon, Ohio.

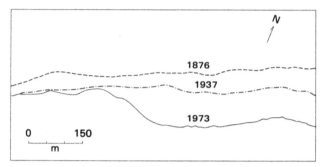

**Figure 11** Recession lines at Painesville-on-the-Lake, Ohio.

**Figure 12** Vertical aerial photograph taken in 1973 at Painesville-on-the-Lake, Ohio.

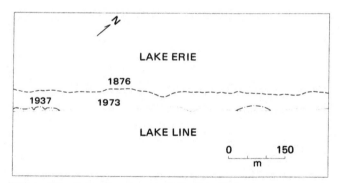

**Figure 13** Recession lines at Lakeline, Ohio.

## DISCUSSION

The principal geomorphic change along the Ohio shore of Lake Erie from 1876–77 to 1973 has been the small (stretches of several tens of meters) to intermediate (stretches of up to 4 km) changes in shape. In general, the overall outline of the shore at a large scale (1:4800) has changed from a relatively smooth, uniform shape to a more irregular, nonuniform shape. Associated with the changes in shape have been the overall decrease in beach width as well as the increase in the number of stretches without beaches. These changes have been caused by the man-made structures.

The seawalls, by armoring short stretches, have generally reduced or eliminated recession behind them, whereas on either side recession has generally continued at the prestructure rate, thus creating a more irregular shore. The groins and jetties, by trapping sand, have generally caused a decrease in recession on the updrift side and an increase in recession on the downdrift side, thus creating a more irregular shore. In general, the longer the structure, the greater the effect; for the large Ohio structures Hartley (1964, p. 30) estimated that "the length of eroding shore is ordinarily five or more times the length of shore which is protected by build-up." We estimate that the Fairport Harbor structures have affected the downdrift (east) shore for at least 4 km. Because of the erosion differential the overall effect of the harbor structures has been to increase recession rates; however, in spite of the marked effect of the harbor structures, there has been a net decrease in recession rates from the early to the late period. We attribute the net decrease in recession rates to the shore-protection structures. These structures have protected enough of the shore to offset the effect of the harbor structures. Moreover, there are dense concentrations of the shore-protection structures downdrift of some of the harbor structures as at Vermilion and Huron.

This decrease in recession rates has also taken place during higher lake levels. Lake level, a most important erosion factor, was greater in 1952, 1972, and 1973 than in any of the years in the early period and the long-term mean was exceeded by 15 cm or more in 11 years in the early period (about 62 years) and 10 years in the late period (about 35 years). Thus if the wave climate was about the same in the two periods, it is likely that a greater amount of wave energy reached the shore per unit of time in the late period; this should have caused an increase in the recession rates of the late period, not the decrease that we have measured, which implies an even greater effect by the shore-protection structures. In general, the changes in shape are least pronounced along the stretches protected by a large number of structures. Here, the changes are smaller and the rates generally decrease with time along the unprotected segments as they become more embayed. Along unprotected stretches fronted by a jetty or a long groin the change in shape can be described as a headland-embayment couple with the stretch updrift of the structure(s) constituting the headland (e.g., Fig. 12).

The decrease in recession rates parallels the decrease in beach width and the increase in number of stretches without beaches. Along this shore the principal source of sand in the beach and nearshore zones is the shore and nearshore. Therefore, erosion of the shore and nearshore is necessary to replace the sand transported by waves and wave-generated currents. The decrease in recession rates has reduced the quantity of sand entering the system and in turn has led to the decrease in beach widths. The harbor structures, by impounding tremendous quantities of sand, also have had a major effect. A comparison of beach widths in 1876–77 and in 1968 downdrift of the Fairport Harbor jetties shows an overall decrease in beach width for at least 4 km. In addition, the small decreases in beach width from 1876–77 to 1968 that we have been able to measure are deceptive with respect to sand volumes. Because the nearshore surface on which the sand rests slopes lakeward, a small decrease in beach width reflects a more than linear loss in sand volume. The greater number of stretches without beaches in 1968 is a reflection of both the decrease in sand supply caused by a decrease in recession as well as the larger number of groins, which, by trapping sand, have caused the same to be distributed more unevenly along the shore. Ironically, the decrease in recession rates has taken place in spite of a decrease in beaches and beach width; that is, the structures have apparently more than compensated for the loss of the best natural form of shore protection – a beach.

## CONCLUSIONS

### Setting, processes, and changes

The 300-km-long Ohio mainland shore of Lake Erie is characterized by gentle (about ½°) nearshore slopes, narrow (< 15 m wide) discontinuous beaches, and an easily eroded (47% till, 26% rock, 22% sand, and 5% laminated clay) shore. Storm waves up to 1 m in height combined with wind setups of 1 or 2 m do the bulk of the geomorphic work. Wave erosion is greatly affected by long-term lake-level fluctuations, which range up to nearly 2 m, as well as by ice, which usually covers at least the nearshore zone from mid-December to mid-March. Erosion rates appear to follow Sunamura's (1977) expressions:

$$\frac{d_x}{d_t} \propto F \qquad \ln \left( \frac{f_w}{f_r} \right) \propto F$$

Man-made structures have markedly changed the shore. The harbor structures, because of their size, have had significant effects for stretches kilometers in length proximal to the harbors, whereas the shore-protection structures, which have increased in number from about 60 in 1876–77 to about 3600 in 1973, have had significant effects on shorter shoreline stretches along most of the shore. The jetties and groins, by trapping sand and thus reducing the wave energy reaching the shore, commonly have helped protect the updrift shore; simultaneously, by reducing the quantity of sand transported alongshore and thus increasing the wave energy reaching the shore, they have harmed the downdrift shore. The seawalls, by directly armoring the shore from waves, have protected the shore behind them.

Recession overall has decreased from the early (1876–77 to 1937–40) period to the late (1937–40 to 1973) period. For example, 171 km of shore receded at less than 0:3 m/yr in the later period, whereas 151 km of shore receded at this rate in the early period.

Beach widths, like recession, have decreased with time. For example, in 1876–77, 64 km of shore was fronted by wide (≥ 15 m) beaches, whereas in 1968 – at the same or a lower lake level – 35 km of shore was fronted by wide beaches. The beaches have also become more discontinuous; in 1876–77 there were 61 stretches without beaches, whereas in 1968 there were 340.

### Inferences and future implications

The man-made structures have clearly caused changes in the recession rates and in the widths and distribution of the beaches. In general, the shore at a large scale has become more irregular as the structures amplify the differences in recession rates along protected stretches and unprotected stretches. Overall, the increased development of the shore has been accompanied by a tremendous increase in the number of man-made structures –

mostly shore-protection structures – leading to a decrease in erosion. This decrease in erosion in turn has led to a reduced sand supply – the shore is a major source of sand for the beaches – and thus narrower beaches.

If present trends continue, further development of the shore seems likely. We foresee increased structural protection and subsequent decreases in erosion and beach widths. Because beaches are valuable shore and marsh protection and provide recreational space as well as a coastal environment, what can be done to improve the situation? Assuming an increase in the number of shore-protection structures, perhaps the most viable alternative is beach nourishment. Major harbor jetties have trapped large quantities of sand and there are sizable offshore sand deposits (Hartley 1960) off Fairport Harbor ($\sim$320 $\times$ 10$^6$ m$^3$) and Lorain-Vermilion ($\sim$100 $\times$ 10$^6$ m$^3$), so that sand supply does not appear to be a problem. Moreover, the more irregular shore may help reduce the longshore flux of sand, thus reducing the frequency of nourishment. Even today, the pocket beaches in small embayments appear relatively stable with respect to the existing wave climate and high lake levels. Thus the final decision may rest on economic or political grounds.

## ACKNOWLEDGMENTS

Many people contributed to the county studies. W.R. Lemke and D.L. Liebenthal helped throughout, particularly in the field. Valuable field help was rendered in the summer of 1973 by Terry Van Offeren, and in the summers of 1974 and 1976 by T.J. Feldkamp; Van Offeren and Feldkamp also assisted in the office. Merrianne Hackathorn did the editing, V.J. Saylor prepared the final figures, and M.S. Longer typed the manuscript. We thank H.R. Collins, State Geologist, for permission to publish this paper.

## REFERENCES

Bajorunas, L. 1961. Littoral transport in the Great Lakes. *Proc. 7th Coastal Eng. Conf.,* The Hague, Netherlands, 1960, 326-41.

Benson, D.J. 1978. *Lake Erie shore erosion and flooding, Lucas County, Ohio.* Ohio Geol. Survey Rept. Invest. 107.

Carter, C.H. 1976a. Recession rates of rock, till and glaciolacustrine clay along the U.S. shore of Lake Erie. *North-Central Sec. Geol. Soc. America Mtg., Kalamazoo, Mich.,* 470.

Carter, C.H. 1976b. *Lake Erie shore erosion, Lake County, Ohio: setting, processes, and recession rates from 1876 to 1973.* Ohio Geol. Survey Rept. Invest. 99.

Carter, C.H. and D.E. Guy, Jr. 1980. *Lake Erie shore erosion and flooding, Erie and Sandusky counties, Ohio: setting, processes, and recession rates from 1877 to 1973.* Ohio Geol. Survey Rept. Invest. 115.

Environmental Data and Information Service 1978. *Local climatological data (annual summary, Cleveland, Ohio).* U.S. Department of Commerce, National Climatic Center.

Hartley, R.P. 1960. *Sand dredging areas in Lake Erie.* Ohio Div. Shore Erosion Tech. Rept. 5.

Hartley, R.P. 1964. *Effects of large structures on the Ohio shore of Lake Erie.* Ohio Geol. Survey Rept. Invest. 53.

Lake Survey Center 1973. *Monthly bulletin of lake levels.* U.S. Department of Commerce, NOAA–National Ocean Survey, 1 sheet.

Saville, T., Jr. 1953. *Wave and lake level statistics for Lake Erie.* U.S. Army, Corps of Engineers, Beach Erosion Board Tech. Memo 37.

Sunamura, T. 1977. A relationship between wave-induced cliff erosion and erosive force of waves. *Jour. Geology* **85**, 613–8.

U.S. Army, Chief of Engineers 1932. Preliminary examination and survey of Fairport Harbor, Ohio. *House Doc. 472, U.S. 72nd Cong., 2nd sess.*

U.S. Army, Corps of Engineers 1950. Appendix IX: Shore of Lake Erie in Lake County, Ohio, beach erosion control study. *House Doc. 596, U.S. 81st Cong., 2nd sess.*

U.S. Weather Bureau 1959. *Climatology and weather services of the St. Lawrence Seaway and Great Lakes.* Tech. Paper 35.

Whittlesey, C. 1838. *Geological report.* Ohio Geol. Survey 2nd Ann. Rept., 41–71.

# 11

# EROSION HAZARDS ALONG THE
# MID-ATLANTIC COAST

*Robert Dolan, Bruce Hayden, Suzette K. May, and Paul May*

## ABSTRACT

Through analyses of historical changes in the shoreline and overwash zone (storm–surge penetration zone) along 630 km of the mid-Atlantic coast, we have determined that there are along-the-coast patterns of the rates of change of the shoreline. These arcuate patterns are present both before and after the passage of severe storms and are persistent through time. Severe storms do not result in a total restructuring of the shore zone. Like most natural systems, coastline processes and coastline forms are organized in both time and space, so high hazard zones are predictable from a probabilistic standpoint. Based on our studies this predictability extends down in scale to just a few hundreds of meters along the coast.

## INTRODUCTION

Hurricanes and severe winter storms are responsible for frequent shore-zone changes along the mid-Atlantic coast barrier islands. In addition to physical and ecological modifications, private land holdings are often destroyed, communication and transportation facilities are disrupted, and loss of life is not uncommon (Fig. 1). Storm damage is superimposed on the longer-term trends of erosion and accretion that are related to sea-level variations (Hicks & Crosby 1975), the supply of sediment to the coast (Wolman 1971), secular changes in wave climate (Hayden 1975), and human alterations of the shoreline (Dolan 1972). In spite of an obvious long-term trend of shoreline recession and the effects of periodic storms, with few exceptions coastal zone planning and development have been based on the concept that the shoreline is stable or that it can be engineered to remain stable.

Understanding the natural dynamics of barrier islands is essential to recognizing and estimating both the short-term and long-term hazards of living on them. The purpose of this paper is to summarize the current information on barrier island dynamics and to show relationships between these dynamics and environmental hazards.

## BARRIER ISLAND FORMATION

The mechanics of barrier island formation and migration has been a subject of debate among geomorphologists for many years (Shepard 1960; Hoyt 1967; Hoyt & Henry 1967; Swift 1968 1975; Otvos 1970; Schwartz 1971). There is, however, indisputable

**Figure 1** Cape Hatteras, North Carolina. The Lincoln's Birthday storm of 1973 destroyed the protective barrier dune system and damaged the seaward line of development. (Photo by R. Dolan.)

evidence that the mid-Atlantic barrier islands are migrating landward (Kraft et al. 1976; Fisher & Simpson 1979). Peat deposits and tree stumps, remnants of forest stands on the back sides of the islands, now are being found on open ocean beaches, indicating distinct marine transgressions (Curray 1960; Kraft et al. 1973; Dillon & Oldale 1978; Field et al. 1979), and recession rates can be measured easily from historical maps and aerial photographs. These changes are a function of three factors: the amount of sediment within a coastal segment, the magnitude of natural processes, and the stability of sea level – all factors directly related to the geological origin of the barrier islands.

Sea level was approximately 120 m lower 12 000 to 15 000 years ago, and the shoreline of the Atlantic coast was 60 to 150 km seaward of the present position. With the change from the Wisconsin glacial stage to the current interglacial stage, the sea started to rise and continued to rise for 8000 years (Curray 1961; Coleman & Smith 1964), reaching the present level about 4000 or 5000 years ago.

As the sea rose and the shoreline was driven across the continental shelf, large masses of sand were moved with the migrating shore zone in the form of beach deposits (Duane et al. 1972; Field & Duane 1976). Sediment that had been deposited as deltas within the coastal river systems was also reworked by wave action and moved along the shore.

Once sea level became fairly stable, waves, currents, and winds worked together on the sand to form the beaches and barrier islands that now stretch from New England to Texas (Fig. 2). As long as the inshore system contained surplus sediment, the beaches continued to build seaward until equilibrium was established. Equilibrium in this case was a function of the balance among storm and wave energy, sea level, and the amount of sediment in the transport system.

Years B.P.

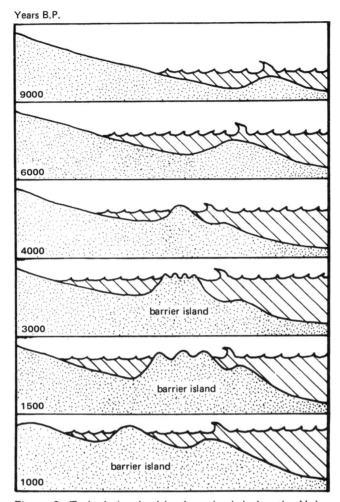

**Figure 2** Today's barrier islands evolved during the Holocene as the sea level gradually increased. During the early
Holocene we believe there was a period when surpluses of
sediment were available to be reworked by shoreline processes to form the barrier islands we see today. In recent
decades there seems to be much less sand available, and the
islands are eroding and moving landward. (Credit: *American
Scientist.*)

Although sea level remained fairly stable following the post-Wisconsin rise, sea level
has been rising slowly for the past 2000 years. This slow adjustment resulted in the
recession of shorelines and the enlargement of bays and sounds. Over the past 100 years,
the rise has been more rapid, totaling slightly more than 30 cm (Hicks and Crosby 1975).
The rate of barrier island recession over the last 2000 years undoubtedly varied as the rate
of the rise of sea level changed, as the supply of sand waned, and as the slope of the
bottom of the inshore zone evolved in response to storms and waves.

**Figure 3** The Great Atlantic Coast Storm of March 7, 1962, generated a 2-m storm surge and 5- to 7-m waves. The results: dunes destroyed and extensive overwashing. Along the Outer Banks of North Carolina the zone of overwashed sediment (heavy line) averaged 250 m wide. (Photo by R. Dolan.)

Overwash and inlet formation are the two most important processes in the landward movement of the barrier islands. During severe storms, the beach zone and seaward dunes are overtopped by waves coupled with high water levels (Fig. 3). As this sediment-charged mass of water spills across the beach and flows toward the bays and sounds on the inland margins of the islands, a layer of sediment is removed from the beach and added to the island's interior – a process that transforms the shape and position of the island but conserves its total sediment mass (Godfrey 1970; Dolan et al. 1973).

## MEASUREMENT OF SHORELINE DYNAMICS

Analysis of historical changes in the shoreline and overwash zone requires repeated sampling in space and time. Although information of this type can be obtained from ground surveys, maps, and charts, our research has led us to believe that aerial photography is the most reliable source for high-resolution, regional-scale data of shoreline change along the barrier islands. Therefore, in 1975 we began development of a common-scale mapping system to provide a uniform data base for both intra- and inter-barrier island comparisons (Dolan et al. 1978a).

Using the common-scale mapping system, we have produced and analyzed data on shoreline and storm-surge penetration changes (Fig. 4) and rates of change at 100-m intervals along 630 km of the Atlantic shoreline between New Jersey and North Carolina. Figure 5 shows the rates of shoreline change for the 6300 transects of the mid-Atlantic reach. The calculation of mean rates of change for entire barrier islands does not give an

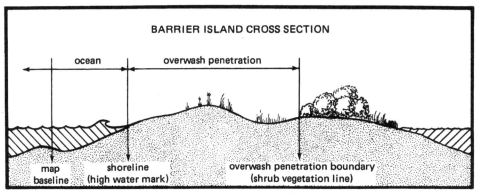

**Figure 4**   Shoreline and storm-surge penetration line. (R. Dolan.)

accurate picture of the dynamics, as most of the islands have higher rates of change at one end than at the other. If the mean value is used, however, the overall shoreline erosion rate for this section of the mid-Atlantic coast is 1·5 m/yr. Islands with more southerly exposures have lower rates of erosion and some may even be accreting. These rates are a function of the direction of storm tracks and wave approach and of the orientation of the shoreline along the mid-Atlantic coast (Dolan et al. 1979b).

## Prediction

A primary use of our common-scale shoreline data base is the prediction of future positions of the shoreline and the landward limits of surge damage for specified time intervals and probability levels. Data required for these calculations are the average rates of change of the shoreline, which are normally distributed (Dolan et al. 1978b) and storm-surge penetration line with associated standard deviations to indicate an expected variability. The landward limit ($\Delta S$) of the shoreline for a given time interval ($\Delta t$) and for a given probability level ($p$) is given by Dolan et al. (1978b) as $\Delta S = \Delta t [(\text{rate of change}) + k (\text{standard deviation})]$, where $k$ is the number of standard deviations required to give the desired probability level. A similar equation will give the projected landward limit of the storm-surge penetration line. These predictions, of course, assume that the trends of the past 30 to 40 years will continue essentially the same in the future.

Table 1 lists 20- and 50-year predictions of the shoreline position for the mid-Atlantic coast. Our predictions are made, using the 1980 shoreline position as the datum, at probability levels of 84% ($p$ = 0·84, $k$ = 1) and 50% ($p$ = 0·5, $k$ = 0). Table 2 lists the landward limit of the overwash penetration (storm-surge) line at the same probability levels. Values are in meters landward (-) or seaward (+) of the present 1980 positions. A given value indicates that there is an 84% (or 50%) probability that the shoreline and overwash penetration limit positions will change by that amount or more over the time period indicated. Armed with this information, planners can devise hazard zones based on the level of risk that land developers are willing to assume.

Whereas the mean rate of shoreline change for 630 km of the mid-Atlantic coast is -1·5 m/yr (erosion), some areas have significantly higher rates and are readily recognized as hazardous places for development. In areas with high erosion rates, such as along the northern part of Assateague Island, Maryland, and at Drum Inlet on the Outer Banks of

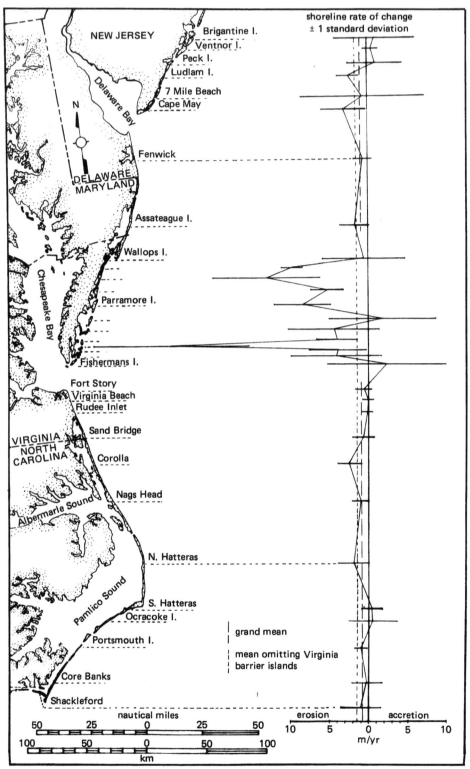

**Figure 5** Mean rate of shoreline change (erosion/accretion) in m/yr for a 670-km reach of Atlantic coast.

Table 1   Twenty- and 50-year predictions of the shoreline position[a]

| | Shoreline position (m) from 1980 location[b] | | | |
| | 84% | | 50% | |
| Location | 20 | 50 | 20 | 50 |
|---|---|---|---|---|
| Brigantine, N J | −52·1 | −136·0 | −56·0 | −140·0 |
| Atlantic City, N J | 35·6 | 83·6 | 32·0 | 80·0 |
| Ocean City, N J | −21·0 | −60·0 | −26·0 | −65·0 |
| Avalon, N J | 25·2 | 55·2 | 20·0 | 50·0 |
| Cape May, N J | 4·4 | 4·4 | 0·0 | 0·0 |
| Cape Henlopen, DE | 116·0 | 272·0 | 104·0 | 260·0 |
| Rehoboth, DE | 20·3 | 44·3 | 16·0 | 40·0 |
| Fenwick Light, DE | −9·4 | −27·4 | −12·0 | −30·0 |
| Ocean City, MD | 32·4 | 77·4 | 30·0 | 75·0 |
| Assateague Island (north) | −176·9 | −458·9 | −188·0 | −470·0 |
| Fishing Point, VA | c | c | 504·0 | 1260·0 |
| Cape Henry, VA | −18·0 | −48·0 | −20·0 | −50·0 |
| Virginia Beach, VA | −2·0 | −8·0 | −4·0 | −10·0 |
| Duck, NC | −13·7 | −40·7 | −18·0 | −45·0 |
| Jeannette's Pier, NC | 27·4 | 63·4 | 24·0 | 60·0 |
| Salvo, NC | 57·1 | 114·1 | 38·0 | 95·0 |
| Avon, NC | −13·4 | −43·4 | −20·0 | −50·0 |
| Cape Hatteras Light | −11·1 | −32·1 | −14·0 | −35·0 |
| Ocracoke Light | 74·1 | 152·1 | 52·0 | 130·0 |
| Portsmouth, NC | −13·0 | −8·0 | −14·0 | −35·0 |
| Drum Inlet Coast Guard St., NC | −148·4 | −358·4 | −158·0 | −395·0 |
| Cape Lookout, NC | 21·5 | 48·5 | 18·0 | 45·0 |

[a] Datum is 1980 shoreline position.
[b] Minus sign indicates erosion.
[c] Information not available.

North Carolina, the predicted shoreline position exceeds the 1980 boundary for the islands' widths.

## Barrier island configuration and shore-zone hazards

The Atlantic coast barrier islands have crescent-shaped shoreline patterns that are the result of variations in shoreline processes. For example, the rate of shoreline erosion on the Virginia barrier islands varies systematically with the configuration of the shoreline, with erosion rates greatest where the azimuths of the shoreline are near 28° east of north, and less at smaller and larger azimuth angles (Dolan et al. 1979b). At places the shoreline is concave and at other locations, convex.

The largest of the crescentic forms that occur are within the large arcs of the Carolina capes. Smaller forms include beach cusps (10 to 30 m), giant beach cusps (100 to 200 m), and larger forms up to 1 km or more in wavelength (Dolan et al. 1974). Inshore bars and troughs also assume crescentic and rhythmic configurations in response to sea states, tides, and sea level.

These crescentic forms must result from differential and periodic shore-zone processes. In earlier papers we showed that along-the-coast periodicities exist in shoreline

**Table 2**  Twenty- and 50-year predictions of the overwash penetration limit[a]

| Location | Overwash penetration limit (m) from present location[b] | | | |
|---|---|---|---|---|
| | 84% | | 50% | |
| | 20 | 50 | 20 | 50 |
| Brigantine, NJ | 21·1 | −0·9 | −2·0 | −5·0 |
| Atlantic City, NJ | 234·6 | 528·6 | 196·0 | 490·0 |
| Ocean City, NJ | −13·6 | −49·6 | −24·0 | −60·0 |
| Avalon, NJ | 5·8 | 5·8 | 0·0 | 0·0 |
| Cape May, NJ | 18·7 | 18·7 | 0·0 | 0·0 |
| Cape Henlopen, DE | −170·9 | −434·9 | −176·0 | −440·0 |
| Rehoboth, DE | 3·3 | 3·3 | 0·0 | 0·0 |
| Fenwick Light, DE | −30·6 | −96·6 | −44·0 | −110·0 |
| Ocean City, MD | 32·0 | 74·0 | 28·0 | 70·0 |
| Assateague Island (north) | 112·0 | 274·0 | 108·0 | 270·0 |
| Fishing Point, VA | c | c | −644·0 | −1010·0 |
| Cape Henry, VA | 40·0 | 88·0 | 32·0 | 80·0 |
| Virginia Beach, VA | −33·9 | −96·9 | −42·0 | −105·0 |
| Duck, NC | −349·1 | −907·1 | −372·0 | −930·0 |
| Jeannette's Pier, NC | 34·3 | 82·3 | 32·0 | 80·0 |
| Salvo, NC | −104·7 | −287·7 | −122·0 | −305·0 |
| Avon, NC | −89·2 | −230·2 | −94·0 | −235·0 |
| Cape Hatteras Light | 6·4 | 6·4 | 0·0 | 0·0 |
| Ocracoke Light | 64·4 | 130·8 | 44·0 | 110·0 |
| Portsmouth, NC | c | c | c | c |
| Drum Inlet Coast Guard St., NC | −84·3 | −228·3 | −96·0 | −240·0 |
| Cape Lookout, NC | −126·5 | −339·5 | −142·0 | −355·0 |

[a] 1980 datum.
[b] Minus sign indicates decreasing overwash zone.
[c] Information not available.

rates of change at scales ranging from kilometers to tens of kilometers based on analyses of rates of shoreline change (recession or progradation) and the landward limit of the storm-surge penetration (Dolan et al. 1979a, 1979b; Hayden et al. 1979). We have suggested that these periodicities may be associated with standing waves or a related process, although at this point no one can make a definitive statement on the process–response relationship.

The pattern of storm-surge deposits along the Delaware coast following the Ash Wednesday storm of 1962 is shown in Figure 6. Although all sections showed some degree of overwash, the distance that sand penetrated inland varied along the coast. Spectral analysis of these patterns and the rates of shoreline change over the last 40 years suggest that along-the-coast periodicities (regular, recurring patterns) exist for both the long-term average shoreline erosion and the penetration during a single storm. We are convinced that whatever process is causing these periodicities plays an important role in determining the spatial pattern of erosion and storm damage (Dolan et al. 1979a), and that there is a natural template of change for all sedimentary coasts.

**Figure 6**  Patterns of sediment deposition resulting from storm-surge overwash following the Ash Wednesday storm of 1962 along the Delaware coast. Arrows indicate individual overwash deposits. Analysis of along-the-coast shoreline erosion patterns indicates periodic variations of similar wavelength. (Photo courtesy of U.S. Air Force.)

## Analysis of spatial patterns

The fundamental question in our research is whether or not the arcuate patterns, and thus the assumed forcing processes, are systematically distributed along the coast. The spectral analysis we used to answer this question followed the procedures outlined by Rayner (1971). This method was modified by the application of a 95% confidence interval factor (CIF), as calculated from the $\chi^2$ distribution:

$$CIF = \chi[\nu, 0.95]$$

where $\nu = 2N/(N/4)$ (using $M = N/4$ lags) degrees of freedom (dof). Adjustments are made for the end points of the spectra where only $\nu/2$ dof exist. The 95% confidence limit (dashed line) is then determined by multiplying the low-resolution spectra, computed using $m = N/20$ lags, by the calculated CIF. Therefore, the probability that the true value lies below this limit is 95%, whereas those peaks that extend above the limit may be considered significant (Bath 1974).

Table 3 lists the wavelengths of those peaks that are significant at the 95% level. For those reaches where no peaks are significant at that level, the peaks encompassing the most power are listed. Figure 7 shows the plots of the spectra of the shoreline rate of change and the overwash penetration distance for several distinct reaches of the mid-Atlantic coast. Comparison of the SLX and OPDX spectra for each reach shows a correlation between the patterns of shoreline change and the degree of overwash penetration. These along-the-coast patterns are independent of isolated extreme events (storms) and are persistent through time (Dolan & Hayden 1981).

The implications of this are clear. Like most natural systems, coastline processes are organized both in time and space. Therefore, positions along the coast with histories of high shoreline mobility will continue to be hazardous places for development.

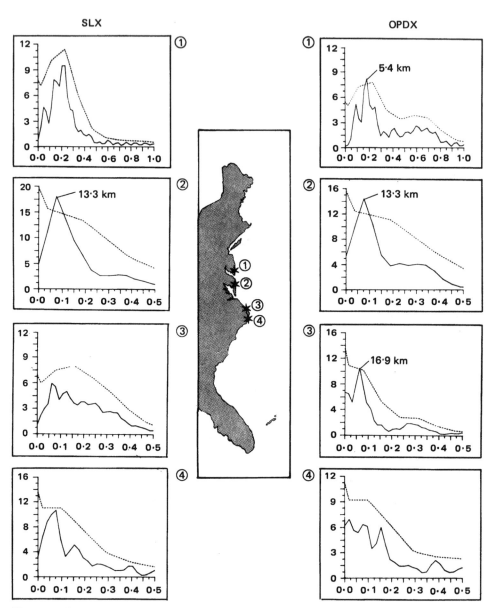

**Figure 7**  Power spectra of the rates of change of shoreline position (SLX) and the limit of overwash penetration (OPDX) for each of the four test sites along the mid-Atlantic coast are shown. Spectral peaks significant at the 95% confidence level are labeled on each of the curves. The percent variance explained is given on the ordinate and the frequency in cycles per kilometer on the abcissa.

**Table 3**  Dominant spectral wavelengths for shoreline mean rate of change (SLX) and mean overwash penetration distance (OPDX) for selected reaches of the Mid-Atlantic Coast

| Coastal reach | $N^a$ | SLX (km$^{-1}$) | OPDX (km$^{-1}$) |
|---|---|---|---|
| Southern New Jersey[b] | 875 | 21·8[c] | 10·9[c] |
| | | 7·3[c] | 5·5 |
| | | 4·8[c] | |
| Fenwick Island | 535 | 13·3 | 13·3 |
| Assateague Island | 615 | 15·3[c] | 30·6[c] |
| | | | 6·1[c] |
| Virginia Barriers[b] | 1072 | 26·8[c] | 53·6 |
| | | | 26·8[c] |
| Cape Henry – Oregon Inlet | 1267 | 63·2 | 63·2 |
| | | 31·6 | |
| Oregon Inlet – Hatteras[b] | 1285 | 15·9[c] | 16·0 |
| Hatteras – Lookout[b] | 1036 | 12·9[c] | 12·9[c] |
| | | | 6·5[c] |

[a] N, number of consecutive 100-m observation stations included in reach.
[b] Binomial filter of order 8 used.
[c] Not significant at the 95% confidence level.

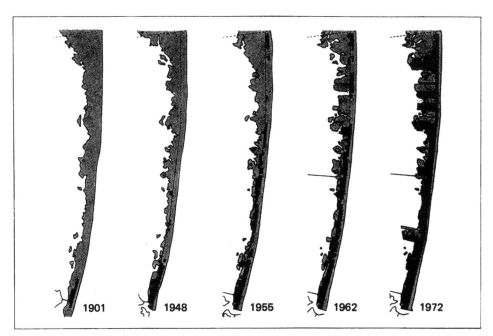

**Figure 8**  Ocean City, Maryland, developed from a small seashore community in 1945 to a large urban complex by 1972. The black shading indicates the areas developed in successive years. It is apparent that little land remains open for further development. (J. Bitting.)

## Patterns of human intervention

Despite the well-established, long-term trend of barrier island migration and the effects of periodic storms, as well as repeated warnings by NOAA, the Department of the Interior, and the U.S. Army Corps of Engineers, coastal zone development has progressed at a rapid rate (Figs. 8 and 9).

Recent trends in development on the Atlantic and Gulf coast barrier islands have been analyzed by the USGS (Lins 1981). Of 282 islands surveyed, roughly 70 are developed or urbanized. About 80 other islands have been purchased for, or included within, state and local recreation areas or preserves. Fifteen of the largest barrier islands have been acquired by the federal government for wildlife refuges and national seashores. The remaining 120 islands are privately owned and largely undeveloped.

The USGS study included analyses of changes in land use and land cover on the 282 islands for the period 1945-73. Four categories – wetland, urban or builtup land, barren land (sand flats on overwash terraces), and maritime forests – account for 90% of the total area of barrier islands. Land used for urban development has increased 56 000 ha, or 53%, during the 30-year period studied. Urban land accounted for only 5.5% of the total of 670 000 ha in 1945; in 1975 it accounted for 14%. Most of this development

**Figure 9**   Most of the sand seaward of the present line of development at Ocean City, Maryland, is gone (left background). Beach nourishment is the only viable solution to maintain the beach front (foreground). (Photo by R. Dolan.)

occurred in wetland areas (32 000 ha) and, to a lesser extent, in forests (6500 ha) and barren lands (11 000 ha).

Despite the rapid expansion of residential and commercial development, the type of land cover on barrier islands in 1975 was still predominantly wetland; barren land occupied another 93 000 ha, or 14%; and maritime forests covered 60 000 ha, or about 10%. It is noteworthy that the area of urban or builtup land equaled the area of barren land. Furthermore, 14% urban area represents a very large percentage in view of the fact that only 3% of the total land area in the United States is urban (Hart 1975).

The potential for catastrophic loss increases with rapid development of barrier islands. In fact, the hazards of coastal erosion and storm surge are not constant but increase annually. The increase stems not from greater exposure of life and property but rather from the dynamics of the islands and, in some cases, from man's attempt to control the natural processes. The islands are, in effect, migrating out from under coastal development, rendering each building, in its turn, the status of more vulnerable beach front property (Fig. 10). Many communities that reduced the risks by constructing barrier dunes, sea walls, groins, and other structures during the 1940s and 1950s now face losing their beaches and incurring rapidly increasing maintenance costs on these protective structures (Fig. 11). Along the Outer Banks of North Carolina, dune breaches and overwashes have become more common each year, and there is little room left on narrow beaches to build more barrier dunes without major adjustments in the property lines and relocation of the highway. We have also found that the patterns of high shoreline mobility are regular along the coast. Areas of dune breaching and deep inland storm-surge penetration are areas that will respond to future storms in a similar manner.

**Figure 10**  The Arlington Hotel at Nags Head, North Carolina, was constructed in the early 1900s about 100 m from the shoreline. In 1974 a severe northeast storm applied the coup de grâce to this fine old establishment. (Photo courtesy of Aycock Brown.)

**Figure 11** Cape May, New Jersey. Although this rock sea wall prevents further recession of the shoreline, the beach continues to erode, virtually in this case, until the subaerial beach disappears. (Photo courtesy of Jeff Heywood.)

The problems of risks and hazards in coastal environments differ from floodplain problems because of the trend term in the shoreline equation. This precludes estimates of storm-surge return intervals for assigning hazard probabilities. Accordingly, given a location on an eroding coast, the hazard due to a "100-year storm" increases systematically with time. Since there may be several causes for shoreline change, the problem cannot be resolved solely by further investigation of storm frequencies and magnitudes. The only alternative is empirical evaluation of historical data. To ensure that an adequate information base is available, a systematic photo reconnaissance and analysis of the coast is needed. As the sample size in erosion studies increases, confidence levels in the derived statistics will improve.

Our investigation also resulted in this final observation. Along the New Jersey coast, in areas that have been engineered to stabilize the shoreline (groins) and to prevent storm-surge penetration (seawalls), the mean rates of change have been greatly reduced – to 1 m or less per year in many areas. However, in these same areas, the standard deviations of the rates of change are high. This suggests that although the engineering works have succeeded in stabilizing the shoreline, when extreme storms do occur, the impact of the storm surge is often greater within the stabilized areas. Thus the hazard of systematic erosion damage is decreased, but the risk of episodic storm damage due to storm-surge penetration is increased.

Time has not changed the nature of the hazards or the problems of developing barrier islands. The dangers from hurricanes and severe winter storms are as great today as they were for the first settlers, and, considering the erosion trend and the large number of people that now live on many of the barrier islands, the risks associated with sedentary occupation are even greater.

## REFERENCES

Bath, M. 1974. Reliability and presentation of spectra. In *Spectral Analysis of Geophysics*, M. Bath (ed.), 193-231. Amsterdam: Elsevier.

Coleman, J.M. and W.G. Smith 1964. Late recent rise of sea level. *Geol. Soc. America Bull.* 75, 833-40.

Curray, J.R. 1960. Sediments and history of Holocene transgression, continental shelf, Northwest Gulf of Mexico. In *Recent sediments, Northwest Gulf of Mexico*, F.P. Shepard et al. (eds.), 221-66. Tulsa, Okla.: American Association of Petroleum Geologists.

Curray, J.R. 1961. Late quaternary sea level – a discussion. *Geol. Soc. America Bull.* 72, 1707-12.

Dillon, W.P. and R.N. Oldale, 1978. Late quaternary sea-level curve. *Geology* 6, 56-60.

Dolan, R. 1972. Barrier dune systems along the Outer Banks of North Carolina, a reappraisal. *Science* 176, 286-8.

Dolan, R. and B. Hayden 1981. Storms and shoreline configuration. *Jour. Sed. Petrology* (in press).

Dolan, R., P.J. Godfrey and W.E. Odum 1973. Man's impact on the barrier islands of North Carolina. *Am. Scientist* 61, 152-62.

Dolan, R., B. Hayden and L. Vincent 1974. Crescentic coastal landforms. *Zeitschr. Geomorphologie* 18(1), 1-12.

Dolan, R., B. Hayden and J.E. Heywood 1978a. A new photogrammetric method for determining shoreline erosion. *Coastal Engineering* 2, 21-39.

Dolan, R., B. Hayden and J.E. Heywood 1978b. Analysis of coastal erosion and storm surge hazards. *Coastal Engineering* 2, 41-53.

Dolan, R., B. Hayden and W. Felder 1979a. Shoreline periodicities and linear offshore shoals. *Jour. Geology* 87, 393-402.

Dolan, R., B. Hayden and C. Jones 1979b. Barrier island configuration. *Science* 204 (4391), 401-3.

Duane, D.B., M.E. Field, E.P. Meisburger et al. 1972. Linear shoals on the Atlantic inner continental shelf, Long Island to Florida. In *Shelf sediment transport, process and pattern*, D.J.P. Swift et al. (eds.), 447-98. Stroudsburg, Pa.: Dowden, Hutchinson & Ross.

Field, M.E. and D.B. Duane 1976. Post-Pleistocene history of the United States inner continental shelf: significance to origin of barrier islands. *Geol. Soc. America Bull.* 87, 691-702.

Field, M.E., E.P. Meisburger, E.A. Stanley and S.J. Williams 1979. Upper Quaternary peat deposits on the Atlantic inner shelf of the United States. *Geol. Soc. America Bull.* 90, 618-28.

Fisher, J.J. and E.J. Simpson 1979. Washover and tidal sedimentation rates as environmental factors in development of a transgressive barrier shoreline. In *Barrier islands*, S.P. Leatherman (ed.), 127-48. New York: Academic Press.

Godfrey, P.J. 1970. *Oceanic overwash and its ecological implications on the Outer Banks of North Carolina*. Washington, D.C.: National Park Service, Office of Natural Science Studies, 37 p.

Hart, J.F. 1975. *The look of the land*. Englewood Cliffs, N.J.: Prentice-Hall.

Hayden, B. 1975. Storm wave climates at Cape Hatteras, North Carolina – recent secular variations. *Science* **190**, 981-3.

Hayden, B., R. Dolan and W. Felder 1979. Spatial and temporal analyses of shoreline variations. *Coastal Engineering* **2**, 351-61.

Hicks, S.D. and J.E. Crosby 1975. *An average long-period sea-level series for the United States*. National Ocean Survey, NOAA Tech. Memo., NOS 15, 6 p.

Hoyt, J.H. 1967. Barrier island formation. *Geol. Soc. America Bull.* **78**, 1125-36.

Hoyt, J.H. and V.J. Henry, Jr. 1967. Influence of island migration on barrier-island sedimentation. *Geol. Soc. America Bull.* **78**, 77-78.

Kraft, J.C., R. Biggs and S. Halsey 1973. Morphology and vertical sedimentary sequence models in Holocene transgressive barrier systems. In *Coastal geomorphology*, D. Coates (ed.), 321-54. Binghamton, N.Y.: State University of New York.

Kraft, J.C., E.A. Allen, D.F. Belknap, D.J. John and E.M. Maurmeyer 1976. *Delaware's changing shoreline*. Dover, Del.: Delaware State Planning Office, 319 p.

Lins, H. 1981. *Patterns and trends of land use and land cover on Atlantic and Gulf Coast barrier islands*. U.S. Geol. Survey Prof. Paper 1156.

Otvos, E.G., Jr. 1970. Development and migration of barrier islands, northern Gulf of Mexico. *Geol. Soc. America Bull.* **81**, 241-6.

Rayner, J.N. 1971. *An introduction to spectral analysis*. London: Pion 174 p.

Schwartz, M.L. 1971. The multiple causality of barrier islands. *Jour. Geology* **79**, 91-94.

Shepard, F.P. 1960. Gulf Coast barriers. In *Recent sediments, Northwest Gulf of Mexico*, F.P. Shepard, F.B. Phleger and T.H. van Andel (eds.), 197-220. Tulsa, Okla.: American Association of Petroleum Geologists.

Swift, D.J.P. 1968. Coastal erosion and transgressive stratigraphy. *Jour. Geology* **76**, 444-56.

Swift, D.J.P. 1975. Barrier island genesis, evidence from the central Atlantic shelf, eastern U.S.A. *Sed. Geology* **14**, 1-43.

Wolman, M.G. 1971. The nation's rivers. *Science* **174**, 905-18.

# 12

# GEOMORPHOLOGY AND LAND SUBSIDENCE IN BANGKOK, THAILAND

## Jon L. Rau and Prinya Nutalaya

### ABSTRACT

The southern Central Plain supports the largest rice-growing area in southeast Asia, covering an area of about 14 000 km². The surface of the plain is covered with a soft marine clay that is so highly compressible that it causes major problems in Bangkok, a city of 5 million. The clay compresses under the load of major structures, roads, and railways because of its low shear strength. Further, intensive groundwater pumpage in Bangkok has resulted in the dewatering of sand aquifers immediately beneath the clay, resulting in a reduction of pore pressure in the clay as water drains from it. It is compressing so rapidly that much of Bangkok will subside below sea level within 10 to 20 years. Field instruments indicate a rate of compression of the upper 50-m section of 3 to 4 cm/yr, corresponding to a rate of decline in the piezometric surface of about 3 to 4 m/yr. The maximum rate of subsidence is more than 10 cm/yr.

## INTRODUCTION

The city of Bangkok, population 5 million (1979), is situated near the mouth of Thailand's main waterway, the Chao Phraya River (Fig. 1). Metropolitan Bangkok, as defined here, is restricted to the Phra Nakorn district on the east side of the Chao Phraya River and some 35 km from the Gulf of Thailand (Fig. 1). Thus the city can roughly be placed at 13°45' north and 100°29' east. Other cities in this latitude are San Salvador, Madras, Manila, and Dakar, and the island of Barbados. The area of greater Bangkok is 300 km². The city services a vast rice-growing area and its waterways handle one-third of the freight moved within Thailand. Thailand's $12 billion (U.S.) gross national product is generated primarily by its farmers, but their produce must pass through Bangkok, where these goods are further refined, packed, and shipped (Anonymous 1980). The country faces new and pressing challenges, such as soaring inflation, hundreds of thousands of refugees, and a military threat from Vietnam on its eastern border.

Another problem is that Bangkok is currently subsiding at a rate of 3 to 4 cm/yr and in 23 years has subsided a maximum of 80 cm. Total subsidence in Bangkok is almost fourteen times the rate of subsidence in Venice (11 cm/43 yr; Berghinz 1976) and continues today. The city is built on a nearly flat plain averaging only 1 m above mean sea level. Beneath the plain a 2000-m section of deltaic clastic sediments rests on a meta-sedimentary–igneous basement. The average elevation of the ground within the city is 1·1 m above mean sea level and its maximum elevation is only 2 m. The average surface

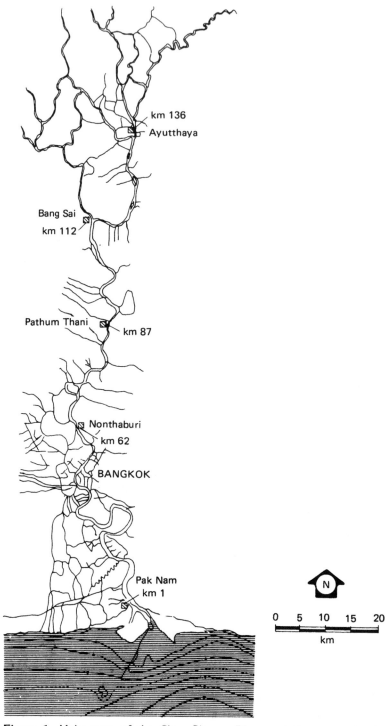

**Figure 1** Main stem of the Chao Phraya River and its estuary in the Bangkok area.

elevation computed from a survey conducted in 1959 (Litchfield, Whiting, Bowne and Associates 1960) should now be less than 1 m because of subsidence.

The surface of the plain in the Bangkok area is covered with a soft marine clay that is compressing due to withdrawal of groundwater. Bangkok expanded so rapidly after 1955 that its city water supply was no longer adequate to handle the needs of its rapidly expanding industrial and agricultural development. In the last 20 years groundwater withdrawal has increased until it totaled 700 000 m$^3$/day in 1974, surging to 940 000 m$^3$/day in 1976. The result was a substantial lowering of the piezometric surface of some of the deeper aquifers (100 to 200 m depth) and a nearly complete dewatering of the shallow aquifers (50 to 75 m depth) in some areas. This was subsequently followed by a decline in pore pressure and partial dewatering of the clays near the top of the section. As a result, flooding is now a more serious problem, especially in the lower-lying areas of the city, where much industry is located and more than 1 million of the city's population live in unauthorized settlements. Further, drainage and sewerage systems malfunction more frequently during the heavy monsoon rains. Well casings are seen protruding above the surface in some parts of the city. Buildings with deep pile supports stand up against the surrounding areas, which continue to subside, resulting in ruptured masonry, sidewalks, steps, and driveways around their perimeters.

Geotechnical studies initiated at the Asian Institute of Technology show that much of the subsidence is related to the dewatering and subsequent compression of a thin soft clay at the surface of the plain. This study was initiated to determine the extent, stratigraphy, and environment of the Bangkok Clay so that a computer model of the areas of subsidence could be used to predict the extent to which subsidence will effect the area in the next 20 years. Moreover, should the government initiate groundwater recharge it is essential to define the geological properties of the clay, which acts as an aquiclude at the surface of the plain.

## Previous investigations

The initial study of Bangkok subsidence problems was by Cox (1968) followed closely by those of Brand and Paveenchana (1971). The most detailed studies of the subsidence problem in terms of its relation to geotechnical properties of the underlying sediments has been by the Division of Geotechnical and Transportation Engineering of the Asian Institute of Technology (1978a, 1978b, 1979, 1980). The principal investigator in these studies is Prinya Nutalaya. Computer simulation of subsidence in Bangkok, based on preliminary data, has been made by Premchitt (1978). No previous studies have been made of the age, extent, and stratigraphy of the Bangkok Clay, although its geotechnical properties are well known. The economic geography of Thailand has been thoroughly reviewed by Wolf (1978), and his work includes references to surface-water and groundwater studies and early newspaper accounts of flooding and subsidence.

# HYDROLOGICAL PROBLEMS

## Flooding

Flooding is still related to particularly heavy periods of monsoon rain, but floods such as occurred in October 1942, when several million dollars damage was done, are no longer possible. Since then the Chao Phraya Dam has been constructed on the main stem of the Chao Phraya in 1957 and the Bhumiphol Dam on one of its major tributaries, the Ping, in 1964. Both have reduced the threat of catastrophic floods at Bangkok. Nevertheless, in the period since 1964 the frequency of moderate flooding has increased.

The geomorphological characteristics of Bangkok are low relief and a river that is barely contained by its natural levees, which rise no more than 1 m above the adjacent back-swamp deposits. Monsoon rains flood the city very rapidly because of their intensity and the inability of the city's drainage system to function efficiently. Subsidence has made the drainage problems more acute. The Chao Phraya can discharge 1500 m$^3$/s below the city without flooding low-lying areas, but during the monsoon season 2000 to 2500 m$^3$/s of water enters the city from the vast areas of the flat plain surrounding the city. Moreover, during the monsoon season the water table rises to the surface and increases the opportunities for sheet flooding and larger discharges in channeled surface runoff. Urban runoff is eventually channeled through tunnels to the Chao Phraya River, where high-pressure pumps lift it to the river. Should heavy rains correspond with high tides and strong onshore winds, the drainage system is not capable of handling the water accumulating on city streets. Major thoroughfares are quickly flooded and traffic within the city is paralyzed. Floodwaters overtop curbs and flood into downtown shops and buildings. In only a few hours rains may flood some streets to a depth of more than 1 m.

Although the city is better prepared now to meet the threat of monsoon flooding, the rapid rate of subsidence may negate the efforts to develop an efficient sewage system. The Thai government is increasingly concerned with the flood threat and the problem of subsidence, although the latter is not taken seriously by the majority of Thai people. Groundwater pumpage, the cause of the subsidence, is uncontrolled at the present time. Until alternative sources of supply are provided there can be no solution to the problem of increasing groundwater pumpage. The present drainage system within the city is operating at a gradient of 1:60 000. The river gradient drops only 25 cm in the 15·5-km river distance through the city. A 30-year project is underway which includes sewage, drainage, and flood protection. Drainage tunnels many kilometers long and 3·3 m in diameter are being constructed. These storm sewers lie at a depth of 4 m and grade to 7·5 m at their lower ends near the great bend of the Chao Phraya. Pumping stations will lift the water 9·5 m and discharge 16 m$^3$/s of water into the river. The new drainage scheme will cost more than $100 million and will drain a 400-km$^2$ area expected to contain 6·5 million inhabitants by 1992. This area will be protected from floods by the installation of a sewer system, aqueducts, the deepening of canals, and the elevation of existing roads. But the subsidence problem remains and becomes more acute. Just what effect it will have on the new drainage system is impossible to predict at this time.

## Groundwater pumpage

Groundwater has been developed for almost 60 years in the Bangkok area. The first municipal well was drilled in 1953 and more than 150 production wells were drilled between 1957 and 1960. In the last 20 years an additional 100 municipal wells have been drilled. The Metropolitan Water Works Authority estimates that city wells produced about 330 000 m$^3$/day in 1976 (Metcalf and Eddy, Inc. 1978). The public wells yielded from 660 to 9600 m$^3$/day. Private wells number more than 6000 and yield from 50 to 9600 m$^3$/day. Private wells supply a variety of users from private homes to hotels. Many government offices, hospitals, schools, and housing developments are served by their own

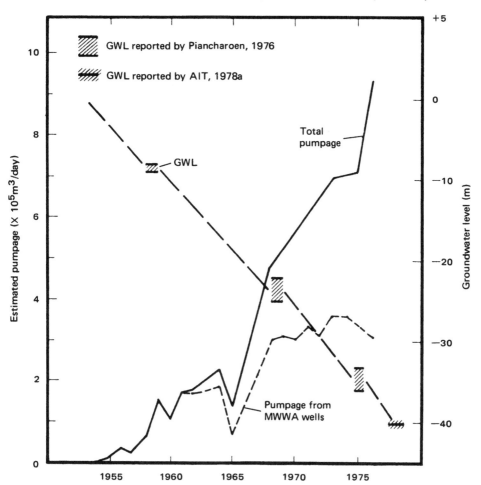

**Figure 2** Estimated pumpage and groundwater level trends in the metropolitan Bangkok area, 1954–1976 (Asian Institute of Technology 1978a).

wells. Many wells have been abandoned because of saltwater intrusion and a rapidly declining groundwater level.

At present, total pumpage probably exceeds 1 000 000 m³/day. As a result two problems have emerged. Before the heavy industrial development and urban expansion of the early 1950s, groundwater levels were within 1 m of the ground surface. By 1959 the piezometric surface dropped to about 9 m below the surface. For the next 20 years the static water level in wells dropped at a rate of 2 to 3 m/year and was 40 m below the ground surface in 1978 (Figs. 2 and 3). Shallow aquifers were dewatered and connate water was induced from the clays sandwiched between the sands in the upper 100 m of section. Some salt water may have been induced through "holes" in the clay layer beneath the Gulf of Thailand, but this has not been proven. Chloride levels in some areas range from 500 to 7000 ppm and are rising rapidly (Metcalf and Eddy, Inc. 1978).

Although the first 200 m of section below the surface contains more than a dozen individual sand lenses, most of the groundwater is derived from the Nakhon Luang

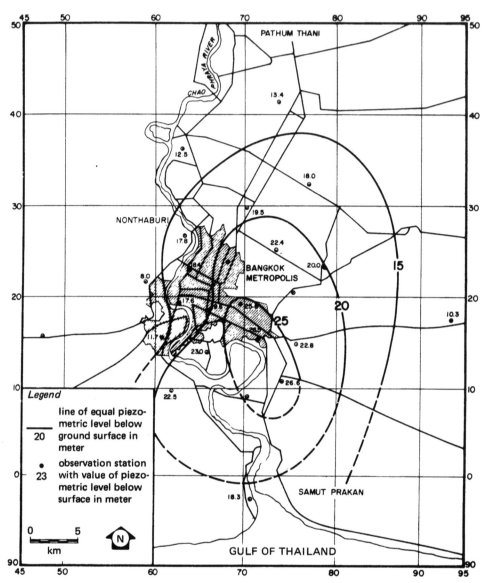

**Figure 3** Contours of piezometric surface below ground level in the Bangkok aquifer (depth 20 to 50 m) based on data collected by the Asian Institute of Technology in October 1978.

aquifer system, a sand and gravel sequence consisting of several interconnected sand bodies at a depth of about 150 m. In the last 2 years the Nonthaburi aquifer system at a depth of 200 m has been increasingly exploited. Most of the wells of the Metropolitan Water Works Authority are developed in the latter. The two most important pumping centers are located in Phra Pradaeng (Fig. 4, grid reference 71/16) and Samrong (Fig. 4, grid reference 75/15). The Samrong pumpage area is close to the area of most rapid subsidence in Bang Na (Fig. 4, grid reference 74/11) and within the north-south-trending

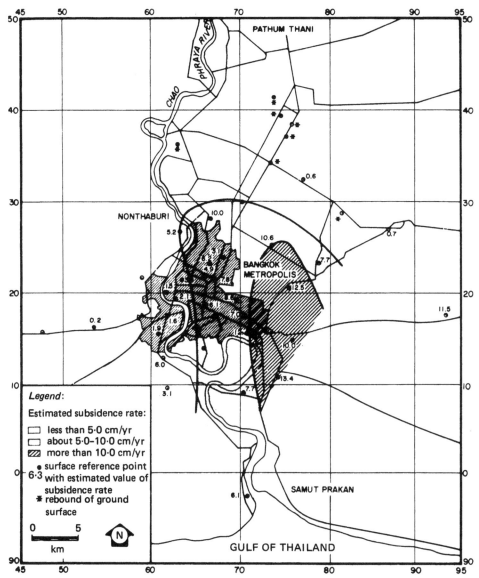

**Figure 4** Surface subsidence rate in cm/yr Bangkok area based on data from two runs of leveling on surface reference points by the Royal Thai Survey Department during the year 1978.

ellipse of maximum subsidence located just southeast of the city center (Fig. 4). Two other major pumping centers are in eastern and northern Bangkok, the latter a major industrial area.

Groundwater levels have declined more than 40 m since 1950. In recent years the groundwater levels in some wells have dropped about 2·5 to 3 m/yr. Is their a direct relationship between the drop in groundwater levels and the compression measured from both surface and subsurface compression indicators?

## Subsidence

Ground surface elevations in Bangkok have subsided by as much as 80 cm over the 28-year period 1950–78. The average rate of subsidence in the area has been 40 to 50 cm over this period and in the last 2 years of precise observations the subsidence rate has been 5 cm/yr, with the maximum rate exceeding 10 cm in a single year in the southeastern part of the city (Fig. 4, grid reference 75/15). Field measurement of land subsidence in Bangkok has been conducted for 3 years. Two independent measurements, surface leveling referred to a stable datum by the Royal Thai Survey Department and measurement on deep compression indicators at a depth of 200 m by the Asian Institute of Technology, recorded a maximum rate of subsidence of 10 cm/yr in the southeastern part of the city. Measurements of compression from both surface and subsurface instruments reveal a zone of subsidence that encompasses all major areas of heavy groundwater pumpage.

The most serious problem resulting from the lowering of the ground surface is the reduction of the hydraulic gradient for drainage and sewage systems. Although the Bangkok situation is not nearly as critical as it was in the past for cities such as Tokyo (Ishii et al. 1976) and Osaka (Yamamoto 1976), where every high-tide season resulted in serious flooding, it is nevertheless extremely serious because of monsoon flooding and the low gradients within the gravity-operated storm sewer system. A subsidence rate of 10 cm/yr can render a drainage system useless and even reverse the gradient throughout a large area.

Bangkok's subsidence zone affects buildings which "stand up" against the surrounding surface if they are founded on deep piles driven to stiff clays or sands. Some buildings supported by deep piles have been connected to others founded on shallow piles, resulting in subsidence of the shallow pile-supported structures and rupturing of the entire structure where the two have been joined together. More commonly the streets and sidewalks go down as the deep pile-supported buildings remain in place. This requires the periodic addition of an extra step in stoops leading into buildings and the repair of the perimeter foundation where it has separated from the sidewalk. Bridge approaches subside while the main rampart of the bridge remains firmly supported on its deep piling. Canals become so sluggish that discharge of their effluent-laden waters to the Chao Phraya takes several days, resulting in vile-smelling areas and toxic waterways in many parts of the city. Groundwater development is also affected by subsidence because it causes damage to well casings and water transmission pipes. Almost all the wells in the Bangkok area protrude above the ground surface (Asian Institute of Technology 1978a). The observed protrusions of wells are larger for wells that have been in service for a long time. Two wells in downtown Bangkok (Lumpini Park) have casings which protrude 50 and 40 cm, respectively, suggesting that subsidence there has been about 0·5 m since their installation 22 to 26 years ago. The casing in a 140-m-deep well at Chulalongkorn University installed 19 years ago today protrudes 35 cm above the subsided ground surface.

*Pattern of subsidence.* The characteristic topographic expression of subsidence is bowl-shaped, with the greatest subsidence at the center of the well field and the total area of subsidence extending some distance beyond the limits of the well field (Dawson 1963). However, the topographic appearance in plain view of the Bangkok subsidence area is the shape of an ellipse with its long axis trending almost north–south parallel to the regional structural grain in the center of the Chao Phraya depression (Fig. 4). The en echelon normal faulting that characterizes the Chao Phraya fault zone may localize the

**Figure 5** Drainage in the lower Central Plain and adjacent areas from LANDSAT imagery and 1:250,000 topographic maps. Compare the drainage pattern of the area underlain by the soft clay (Fig. 9) with that of the adjacent area.

center of subsidence (Figs. 4 and 5). The zone of maximum subsidence is shown on a map by contouring field data using isosubsidence contours for a rate of 10 cm/yr (Fig. 4). The data shown in Figure 4 is the result of leveling conducted by the Royal Thai Survey Department of surface benchmarks. It shows that the most rapidly subsiding area of the

Table 1 Some examples of areas and rates of subsidence

| Location | Area of subsidence[a] (km$^2$) | Period of measurement | Maximum subsidence[b] (cm) |
|---|---|---|---|
| Mexico City | 25 | 1948–60 | 900 (1973) |
| Osaka | 120 | 1948–65 | 300–400 (1965) |
| Tokyo | 230 | 1938–75 | 460 (1975) |
| Taipei Basin | 100 | 1961–75 | 180 |
| Tuscon | 925 | 1948–67 | 320 (1975) |
| Houston–Galveston | 6 475 (1973) | 1943–73 | 230 (1973) |
| San Joaquin Valley | 13 500 (1976) | 1935–66 | 880 (1976) |
| Venice | 10 (1971) | 1930–73 | 11 (1973) |
| Bangkok | 285 (1980) | 1950–80 | 80 (1980) |

[a] Dates are those on which the area of subsidence was measured. The present size of the area of subsidence may be larger or smaller than that indicated.
[b] As of date indicated.

city is located in the Phra Khanong area (Fig. 4, grid point 74/11). Some points within this zone have subsided at a rate of 13·4 cm/yr. The zone of subsidence extends 17 km long and 9 km wide. Its southern limit has not been determined. The area enclosed by a line connecting points of 5 cm/yr of subsidence includes most of metropolitan Bangkok and defines a broader ellipse with a long axis slightly east of south, a width of about 20 km, and a length of more than 33 k. The southern limits of this zone are undefined. The total area included within the subsidence bowl is more than 285 km$^2$. For comparison, the area of subsidence, depth range of compaction (compression), and maximum subsidence of some typical areas are listed below (Table 1).

## METHODS

The methods used to study the subsidence and to relate it to the geology and groundwater pumpage are both geotechnical and geohydrological. During the initial parts of the investigation, twenty-four boreholes up to 50 m in depth were drilled and the soil properties of the sediments penetrated in these wells were measured at 0·5-m intervals. The properties measured included the natural water content, total unit weight, particle size distribution, specific gravity, Atterberg limits, and salt content of the pore water. The coefficient of permeability was measured in the uppermost sand (Bangkok aquifer system). Further, four 100-m holes, four 200-m holes, and one 400-m hole were drilled. Field instruments installed at all stations included compression indicators and piezometers. Compression data were collected from differing intervals from the uppermost 50 m to sections as thick as 400 m. Benchmarks were established at all stations and casings were connected to comparession indicators so that the rate of surface subsidence could be compared to casings cemented into deeper horizons.

Traditionally, geomorphologists have found topographic maps to be their most useful tool, ranking near the top of a list which now includes LANDSAT imagery. Topographic maps have also been valuable in this study, but those portraying flat metropolitan areas must be used with caution. Two scales of topographic maps are available for Thailand, both prepared by the U.S. Army Topographic Command. The larger scale, 1:50 000, is based on information as of 1969 and the other, 1:250 000, is based on information as of 1971. It was the 1:50 000 (Royal Thai map edition 1-RTSD) that was initially considered to be of the most potential value in the study of the geomorphology here. However, recent leveling by the Royal Thai Survey Department has shown that these maps are unreliable for studies of topographic variations of less than 5 m in the lower Central Plain. Spot elevations shown on the Bangkok maps were in error by as much as 4 m. This is a critical error because the Bangkok metropolitan area is only 30 cm to 2 m above mean sea level. Not all of the error can be ascribed to subsidence since the benchmarks were installed. One spot elevation of 6 m located on the Bangkok Metropolis 1:250 000 map (Sheet ND 47-12) across from Klong Toey, the city's main port facility, was actually less than 2 m above mean sea level. However, the maximum subsidence measure in Bangkok is about 80 cm and is located more than 4 km from the Klong Toey docks.

In areas of subsidence there are no permanent benchmarks. All benchmarks are affected by general subsidence and individual benchmarks will show unequal subsidence because of many factors such as rate of groundwater pumpage and the lithology of the compressible units. Subsidence causes serious problems for surveying and mapping, both essential for the determination of elevation and grade in engineering structures. Absolute subsidence must be measured against a fixed reference level such as mean sea level. Leveling with respect to mean sea level must be carried out from a ground reference point that is far removed from the area of subsidence and one that is not likely to subside even if groundwater pumpage were to be initiated near it. The Royal Thai Survey Department selected a bedrock outcrop more than 100 km from Bangkok and all four of its leveling programs were tied into the key benchmark embedded in bedrock. Recently, bedrock was penetrated at a depth of 200 m about 50 km east of Bangkok and future leveling programs may be tied to this bedrock, a gneissic or foliated granite.

## RESULTS

The Chao Phraya Delta begins where four rivers – Ping, Wang, Yom, and Nan – rising in northern Thailand combine to form the Chao Phraya River at Nakhon Sawan, 220 km north of Bangkok (Fig. 5). A fifth tributary, the Pa Sak, draining borderlands on the northeast side of the Central Plain, enters the Chao Phraya near the head of the lower Central Plain at Ayutthaya. For 50 km below Nakhon Sawan the Chao Phraya is a single stream confined by low terraces and scattered hills of bedrock. These bedrock knobs help restrict the flow of the river during the monsoon season, causing sedimentation to occur in the low-lying areas behind them. Below the narrows, in the vicinity of Chainat, the river branches into four distributaries: the Suphan Buri, Noi, Pa Sak, and the Chao Phraya. The completion of the Chainat dam on the main stem of the Chao Phraya in 1957 helped regulate the flow in the distributaries and met irrigation and flood control requirements.

The area of most immediate concern lies below Chainat, where the youngest part of the Chao Phraya Delta is found (Fig. 5). The deltaic area of the Chao Phraya here comprises a roughly rectangular-shaped plain extending from Chainat to the Gulf of Thailand. Its base extends from the Mae Klong River to the Bang Pakong River and includes the Supan Buri or Tha Chin River, draining the west central part of the plain, and the Chao Phraya River in the east central area.

The geologic unit that effects the city of Bangkok, the Bangkok Clay, is restricted to an area extending from the town of Ayutthaya southward to the coast. The elevation of the delta is only 30 cm to 2 m above mean sea level and the slope is very low, about 0·02 in 1000. Predominant features in the deltaic area are the four pairs of natural levees formed by the Chao Phraya and its distributaries. The largest of these follows the Chao Phraya from Chainat to Ayutthaya, a river distance of about 145 km. At its head the levee is about 17 m above mean sea level and at the lower end it is about 2 m. The relative height of the levee with respect to adjacent terrain is only 2 to 3 m. The levee width varies considerably, ranging from 0·5 to 1 km in the upper part and from 1·5 to 2 km in the lower portion. In addition to these four principal levees there are many smaller levees marking the course of former channels. But these are so subtle that they are almost imperceptible to the eye. The Central Plain was covered with an arm of the sea at least as far north as Ayutthaya as recently as 3000 years ago. Little migration of the Chao Phraya has occurred beyond a belt 20 km wide on either side of the present main stem. This may be due partially to the cohesive and clayey texture of the surface of the plain south of Ayutthaya. Early Kings of Thailand ordered the river straightened within the Bangkok area, cutting off several large meander bends, so that now some parts of the straightened channel are without natural levees.

Back swamps occupy the areas between natural levees. Water from rain and floods stagnates in the back-swamp areas for several months of the year; hence the areas are mostly used for paddy fields. The depth of water standing in these areas at the height of the monsoon season ranges up to about 3 m. In the deepest areas the cultivation of floating rice is a well-developed practice. Virtually all of the deltaic area is covered by monsoon floodwaters. This water is essential for the growth of rice because the rainfall alone is inadequate to develop the rice to its full potential. Because of the flat slopes of

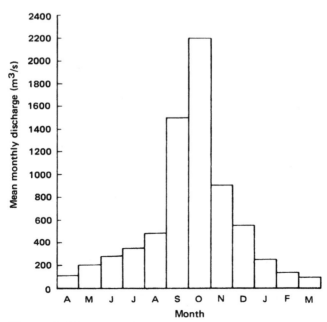

**Figure 6** Mean monthly discharge of the lower Chao Phraya River (1969–1977).

the deltaic area and the comparatively gentle nature of the floods the sediments flooding the area are very fine grained clay size for the most part. The soils are clay rich and heavier than those in other deltas of southeast Asia probably because more clay settles over the surface of this plain than in most other deltaic areas. Thus the monsoon rains are critical for agriculture in Thailand but cause enormous problems for Bangkok, situated directly in the path of the Chao Phraya and fed by more than twenty large canals draining the back-swamp areas adjacent to it.

## Precipitation

The Chao Phraya is fed by rivers that drain mountains on its three sides. Most of the precipitation occurs during the southwest monsoon, when moist streams of air come in from the Bay of Bengal and the Indian Ocean. A dry season occurs from November to April at Bangkok, but the rainy season has two peaks, one when it begins in May and another major peak in September. The average rainfall at Bangkok is 1387 mm, declining slightly to the head of the lower plain, where it reaches 1270 mm at Ayutthaya. The rainfall during the rice-growing season is only 1000 mm and the average requirement for rice is 1500 to 1800 mm. The deficiency is made up by monsoon floods which are carefully directed into the rice paddies at the beginning of the season, and finally swell to their full 3-m depth at the height of season in October (Fig. 6).

## Geology

The setting in which the Bangkok soft clay has been deposited has been termed a depositional coast by Hayes (1964) because the clay was deposited in a coastal zone capping a thick sequence of deltaic sediments. The Quaternary and Tertiary sediments beneath the clay represent a complex sequence totaling more than 2000 m, of which only the uppermost 200 m is well known. Sedimentation in the Central Plain was controlled throughout most of Tertiary and Quaternary time by a combination of tectonic movements both within the plain and in the adjacent mountains. The plain is situated over a large structural depression that has been filled with an assortment of clastic sediments chiefly of fine to medium grain size. This prism of sediment rests upon a basement complex that slopes south to the latitude of Bangkok, where it is broken by an east–west-trending horst block, the Samut Sakhon horst (Figs. 7 and 8). The north-trending axis of the Chao Phraya depression is related to the north–south structural trend of the Paleozoic and Mesozoic fold belt of westernmost Thailand. The sediments reach a thickness of at least 1859 m 15 km west of Bangkok, where a borehole penetrated Upper Cretaceous granite. Thirteen other deep boreholes have been drilled in the lower Central Plain but few have penetrated bedrock. The data suggest that the Chao Phraya trough has been tectonically active during most of Tertiary and Quaternary time, receiving alluvial and deltaic sediments when the adjacent ranges were uplifted. Geophysical evidence suggests that in some areas the sequence of sediments may be more than 4000 m thick. Regional aeromagnetic and seismic data in the Gulf of Thailand indicate that the basement is highly irregular, with relief reflected in basement highs of granitic ridges separated by folded metasedimentary rocks similar in trend and type to those of the southern peninsula of Thailand (Kelley & Reib 1971). Deep-well data from the lower Central Plain also reveal this type of structure (Fig. 8). The geomorphology of the Central Plain is strongly affected by the faulting and most river alignments and strike lines of hills along the margin of the plain are controlled by faults that probably cut the basement. The dominant direction of

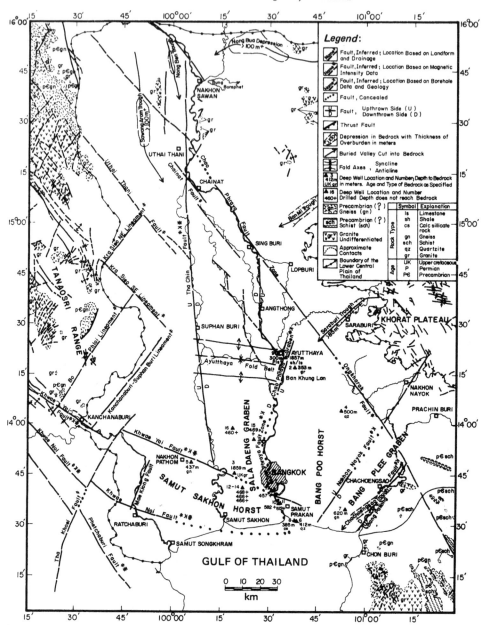

**Figure 7**  Structural framework of the lower Central Plain.

faulting is north–south, although three secondary directions are indicated (Fig. 7). The Chao Phraya River is apparently controlled by en echelon normal faulting indicated by both borehole records and geophysical data in the Ayutthaya area. A 30-km-wide topographic high has been mapped just south of Ayutthaya in the latitude 14°15′ to 14°30′. The soft clay is higher in this east-trending ridge than anywhere else on the plain and may reflect folding of Recent age, although a buried sand body on top of the stiff clay could produce the same landform. Although this belt is only 3 to 4 m above the adjacent plain,

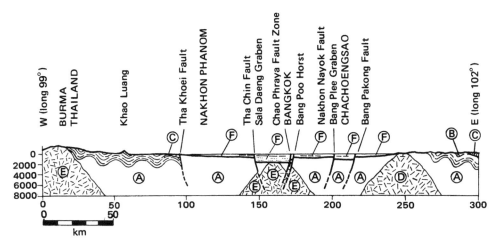

Key to Rock Units:

Ⓐ Precambrian gneiss
Ⓑ Lower Paleozoic sediments
Ⓒ Upper Paleozoic sediments
Ⓓ Triassic granite
Ⓔ Cretaceous granite
Ⓕ Quaternary sediments

**Figure 8**  Geologic cross section along latitude 13°34′ N showing the block-faulted structure beneath the Central Plain.

it controls the drainage in this area, deflecting it either easterly or westerly, and slows the advancing flood of the monsoon season.

## Stratigraphy of the Bangkok Clay

The deposition of marine clay of Holocene age resulted in a low-lying level surface which is the hallmark of the lower Central Plain. The surface of this plain has been modified by riverine processes since the retreat of the last transgression about 3000 years ago. The northernmost limit of this clay is marked by a beach at Bang Sai, where it consists of a remarkably well sorted fine grained sand (Fig. 9). This beach is approximately 5 m above sea level. Only two other unquestionable beach sands have been found at the surface at approximately this altitude. In many places beach sands of this transgression have been buried by reworked terrace and pediment sands and gravels lying immediately adjacent to them. This is especially true of the western side of the plain. But examination of 1:50 000 topographic maps and borehole data, sparse though it may be, have helped confirm that the 5- to 7-m altitude is the approximate limit of the Bangkok Clay. LANDSAT imagery also shows a dramatic change in the drainage pattern and density at this boundary (Fig. 5). Just how long the sea remained in this embayment is unknown, but it is thought that the regression was very slow inasmuch as present-day cities seem to be located along the margin of this old embayment. Historical records indicate that several of these cities were ports as recently as 1500 years ago. Schmidt (n.d.) has suggested that the old capital of the Mon Empire at Lop Buri was a port at this time.

The soft clay unit of the Bangkok Clay as it has been defined by the Asian Institute of Technology (1980) covers an area of approximately 13 800 km² to an average depth of 15 m. It ranges in thickness from 3 to 27 m. The soft clay was deposited on an erosion surface that was moderately dissected prior to the beginning of the transgression about 10 000 years ago. The exact limits of the Bangkok Clay are difficult to determine. The boundary determined from all available borehole data, water well logs, and field work is shown on Figure 9. This transgression is not marked by prominent beaches or shell beds. In the future more data from boreholes will enable the extent of the clay to be mapped with greater accuracy. The present map should be considered a good approximation of its distribution and thickness (Fig. 9).

The term "Bangkok Clay" has been used in several ways. Moh et al. (1969) divided it into three units: (a) weathered clay, (b) soft clay, and (c) stiff clay. "Stiff clay" is a term used by geotechnical engineers to refer to the difference in the shear strength of the clay beneath the soft highly compressible clay at the surface. Buried laterities and a dissected surface at the top of the stiff clay clearly indicate that it is separated from the overlying soft clay by an unconformity. The stiff clay may be several tens of thousands of years older than the soft clay. The weathered clay of Moh et al. (1969) is simply the uppermost meter or two of the Bangkok Clay. Geotechnical data such as the overconsolidation ratio indicate that the crust is consolidated well above unity. Overconsolidation results from desiccation and cementation but also occurs when several tens of meters of clay has been stripped by erosion, exposing a deeper and hence more profoundly consolidated clay than would normally be found at the surface of a normally consolidated but cemented clay. Although the crust began to develop soon after the final regression of the sea from the Central Plain, the total time span available for the weathering of the clay may range from 1500 to zero years, depending on where and if the clay has been exposed to subaerial processes. For example, the soft clay on the floor of the Gulf of Thailand is presumed to be continuous with the blanket of clay now exposed at

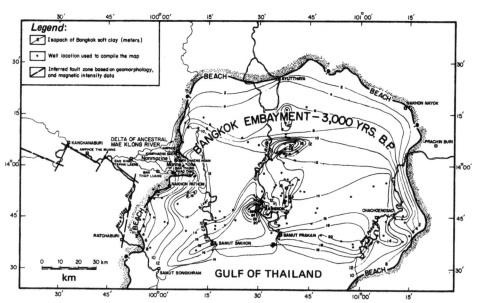

**Figure 9**   Isopach map of the soft unit at the top of the Bangkok Clay.

the surface of the plain, but because the Gulf section has not been exposed to subaerial processes it should have no crust. The presence of organic matter makes the pH values of the crust of the clay acid (Brenner et al. 1978). The pH values approach 4 in the uppermost meter, where organic carbon is as high as 1% (Petersen 1977). At a depth of more than 4 m the mean value of pH increases to about 8·0 (Brenner et al. 1978). The uppermost meter has a low salt content because of the leaching of soluble salt by rain.

The unit beneath the crust is typically a soft olive gray or medium to dark gray clay. It may contain fragments of shells and decayed plant remains. Sand lenses and silt may be interbedded or interlaminated with the clay. Carbonate concretions (calcrete) have been reported from the underlying stiff clay but apparently are absent from the soft clay. Thin beds of peat and organic detritus are common in some areas. The shells are marine bivalves similar to those presently living in the tidal areas of the Gulf of Thailand. The salt content is unusually high and indicates a marine origin. The base of the soft clay rests upon a dissected surface cut into the underlying stiff clay. Laterally, the soft clay may grade into clean, well-sorted fine sand at slightly higher elevations, and these beds have been interpreted as beaches related to the last marine incursion of the Central Plain (Fig. 9). The most abundant clay mineral in the Bangkok Clay is kaolinite, with illite and montmorillonite also present (Petersen 1977). The approximate percentage of kaolinite in the soft clay ranges from 47 to 64%, illite from 10 to 20%, and montmorillonite from 20 to 43%. The silt and sand fraction is mostly quartz with small amounts of feldspar and heavy minerals.

## Extent of the Bangkok Clay

The extent of the soft clay is shown in Figure 9. It can be mapped by using borehole data, topography and LANDSAT imagery. It has never been found above the elevation of 5 to 7 m, as indicated by the 1:50 000 topographic maps. For most of the area shown on Figure 9 it is less than 2 m above mean sea level. If one were able to project the shape of the head of the Gulf of Thailand onto the Central Plain, it is perhaps more than coincidence that the margins of the soft clay are roughly parallel to the width of the head of the Gulf.

It covers an area that is about 145 km wide extending northward from the Gulf inland to Ayutthaya, 90 km to the north. This is an area of roughly 14 000 km². The isopach map reveals that it is not a uniformly thick deposit but varies rather rapidly over short distances, reflecting the uneven surface upon which it was deposited. It is thicker in the Bangkok area and along the north-south-trending Chao Phraya fault zone (Figs. 8 and 9). This may indicate that these faults were active in Holocene time. It thins to less than 10 m in an east-west-trending 35 km wide belt just south of Ayutthaya. The nearly north-south-trending eastern and western margins of the embayment may reflect the normal faulting that parallels the regional structural grain. Several deltas have been built into the Bangkok embayment; these have not been fully deciphered but the two most notable are the Chao Phraya delta in the vicinity of Ayutthaya, where the clay contains much more sand, and the Mae Klong delta on the west in the vicinity of Nakhon Pathom (Fig. 9).

## Age of the Bangkok Clay

In order to estimate the age of the soft upper unit of the Bangkok Clay, eleven radiocarbon dates were obtained from clay deposits near the surface of the Bangkok plain.

Ten of these were taken from the stiff clay underlying the uppermost soft clay and one is from the soft clay itself. The material dated from the soft clay near the Gulf consists of bivalves very similar to those found along the margin of the present Gulf of Thailand. The soft clay near Bangkok at a depth of 15·5 to 16 m is 710 ± 280 years old. This location is about 10 km south of Samut Prakhon (Fig. 9) at the mouth of the Chao Phraya River. Because this sample is from near the base of the clay sequence and presumably was laid down during the transgression, it should be on the order of magnitude of from 6000 to 10 000 years old. The date obtained is irreconcilable with the interpretation presented here.

Other data suggest that the marine clays of southeast Asia are the result of a major late Pleistocene–early Holocene transgression that began in the Gulf of Thailand about 11 170 ± 150 years B.P. (Biswas 1973). This transgression is assumed to have reached the southern margin of the Central Plain by at least 6000 years B.P., as it is well documented that it was standing at +3 m in Perlis (Malaysia) and +6 m in Peninsular Malaysia by 5200 years ± 200 B.P. (Haile 1971). Therefore, the 710-year-old date obtained for the Pom Phrachul deposit seems much too young, especially since it would require more than 16m of subsidence in 700 years. On this basis, the 710 year B.P. date for the soft clay must be rejected. On the other hand, the other ten dates obtained on the stiff clay fall within the range 14 700 to 45 000 B.P. and hence would not exclude an interpretation that places the sea level well below the Central Plain prior to 11 000 B.P. (Biswas 1973). The material dated from the stiff clay consists of limestone nodules (calcrete) that presumably developed at or near the water table in an environment typical of many places in the Central Plain today. Data from Biswas (1973) placing the sea level in the Gulf of Thailand 70 m below its present level at 11 170 ± 150 B.P. seem to further limit the age of the soft clay to less than about 10 000 B.P. This is in agreement with the dates we obtain on the underlying stiff clay, all of which are older than 11 000, the youngest of the collection being 14 700 B.P.

It is clear from Figure 9 that at the height of the Holocene transgression the sea covered most of the Central Plain as far north as Ayutthaya. The beach at Ayutthaya lies about +3 to +5 m above mean sea level and probably correlates with the beach at the same elevation in Peninsular Malaysia, which has been dated as 5200 years B.P. (Haile 1971).

## DISCUSSION

The Bangkok Clay has been found throughout the lower Central Plain over a belt up to 145 km wide. It grades into a beach deposit that probably marks the limit of the Holocene sea in the Central Plain at the time of the last transgression about 3000 B.P. One puzzling piece of data is the historical report that places Lop Buri, the old capital of the Mon Empire located 50 km north of Ayutthaya, on the sea about 1500 years ago (Schmidt n.d.). If this report is true the shoreline at Ayutthaya, placed there after borehole, topographic, and beach sand outcrops were studied, is much too far south. Perhaps Lop Buri was a port on a river leading to the sea some 40 km to the south, just as today Bangkok is 35 km from the mouth of the Chao Phraya.

The environment of deposition of the lower Bangkok Clay is one of a vast tidal flat that has gradually subsided over the period from 45 000 to 14 000 years ago, resulting in a deposit ranging from 10 to 20 m in thickness. The upper part of the Bangkok Clay ranges from 3 to 27 m in thickness and occurs at an altitude of from 0 to 5 m above mean sea level. Inasmuch as the clay is now exposed, either subsidence discontinued sometime in the last 1500 years and the plain was warped slightly upward or sea level was eustatically lowered from 2 to 5 m in the last few thousand years. The former seems more reconcilable with data on eustatic changes from other areas during the Recent. In early Bangkok Clay time (late Pleistocene) the lower Central Plain was a large bay-tidal flat-estuary complex fed by several large rivers and bounded on the south by the Gulf of Thailand. The environment of the mouth of the Chao Phraya River today is a small-scale version of the environment inferred during the late Pleistocene over the width of the Central Plain south of Ayutthaya. The bay that occupied the southern half of the lower Central Plain is referred to here as the Bangkok Embayment (Fig. 9). It extended much farther north during older Pleistocene transgressions. This bay was bordered by strips of coastal salt marsh or mangrove and probably did not receive much river water from the small drainage basins that flank the bay on its east and west. Entering the head of the bay on the northwest was the Tha Chin (Supan Buri) and on the northeast the Pa Sak. Feeding directly into the center of the bay were the ancestral Chao Phraya and Lop Buri Rivers. Most of the coarse debris fed to the Central Plain has been trapped behind the Nakhon Sawan arch, an east–west–trending structural high that has forced the river to cut through a narrows and has slowed the force of monsoon floods across the southern half of the basin. Thus only the finest of sediments make their way into the lower basin. The environment of the upper part or soft section of the Bangkok Clay is interpreted as a delta-front deposit. The evidence for this stems from its molluscan fauna, uniform texture, and abundant slickensides.

## CONCLUSIONS

Geomorphological analysis was used to define the limits of the area of the lower Central Plain underlain by the highly compressible and troublesome Bangkok soft clay. Not only did LANDSAT imagery clearly reveal a difference in the drainage pattern on the clay, but slight differences in topography around the margin of the clay deposit revealed the presence of buried beaches and deltas marking the former limit of the Bangkok transgression. This was later verified by examination of borehole logs in areas of topographic highs around the margin of the embayment. Geomorphology complemented stratigraphic and structural data by revealing the presence of numerous faults down the axis of the Central Plain and trending into it from the mountains of the Thai-Burma foldbelt bordering the plain on the west. Such faults may actually control the areas of subsidence, as they strike parallel to the elliptical trend of the area of maximum subsidence. Finally, geomorphological analysis revealed that flooding in certain sections of the Bangkok area was due not only to subsidence but was also to the low natural levees, the nearly table-top attitude of the back-swamp areas, and the complete absence of natural levees along the man-made reaches constructed by ancient Kings of Thailand. The depth of water in back-swamp areas during the monsoon season and in the urban areas after heavy rain-storms is related to poorly functioning drainage and sewage systems. These systems can be expected to malfunction more frequently in the future as subsidence reduces the natural gradient of storm sewers and canals toward the river, resulting in lower velocity discharges and longer periods of water stagnation within the city.

The problem of subsidence will continue at Bangkok just as long as groundwater pumpage is uncontrolled, in spite of the fact that expensive alterations to the present drainage and sewage systems are being made. The Bangkok soft clay has played a critical role in the subsidence problem for it has been the agent that has taken up much of the compression resulting from the dewatering of the aquifers beneath it. Computer simulation of subsidence has predicted a decline in water levels to a depth of 80 m for the aquifers between 75 and 100 m below land surface and a decline to 100 m for the Nakhon Luang aquifer at a depth of about 150 m. Surface subsidence in the Bang Na area is predicted to be an additional 1·25 m above the present 0·75 m if groundwater pumpage continues unrestrained. These estimates are based on a 5% annual increase in rate of groundwater pumpage, but this may be too low (Asian Institute of Technology 1980).

The Bangkok experience also indicates that geomorphologists should be extremely cautious in using data garnered from elevations on topographic maps in deltaic areas or in Quaternary basins where groundwater pumpage is large, especially if soft clays are present. Many such areas exist in the industrialized cities of Europe and North America and within the great agricultural areas of semi-arid regions of the west. Published data from older reports on the geomorphology of deltaic areas such as land surface profiles and the elevations shown on buried beaches and islands may be irreconcilable with regard to data measured in these areas today. Students of eustatic changes in sea level may arrive at erroneous conclusions should they rely upon data secured in areas of rapid subsidence related to groundwater pumpage in nearby industrial or agricultural centers.

## ACKNOWLEDGMENTS

The authors are especially indebted to Sopit Sodsee, our research associate in the Land Subsidence Project, for her dedication and efforts throughout the three years of the project. Our associates at the Asian Institute of Technology, Dr. R.P. Brenner, Dr. Jerasak Premchitt, and Dr. Tosinobu Akagi, are respectfully acknowledged for their efforts in the analysis of geotechnical data in connection with the Land Subsidence Project, but the authors are solely responsible for any errors in this report.

## REFERENCES

Akagi, T. 1979. Some land subsidence experiences in Japan and their relevance to subsidence in Bangkok, Thailand. *Geotech. Engineering* **10**, 1-48.

Anonymous 1980. Thailand. In *Asia yearbook - 1980*, 290-7. Hong Kong: Far Eastern Economic Review Ltd.

Asian Institute of Technology 1978a. *Investigation of land subsidence caused by deep well pumping in the Bangkok area.* Phase I: *Progress report.* Bangkok: Division of Geotechnical and Transportation Engineering.

Asian Institute of Technology 1978b. *Investigation of land subsidence caused by deep well pumping in the Bangkok area.* Phase I: *Final report.* Bangkok: Division of Geotechnical and Transportation Engineering.

Asian Institute of Technology 1979. *Investigation of land subsidence caused by deep well pumping in the Bangkok area.* Phase II: *Progress report.* Bangkok: Division of Geotechnical and Transportation Engineering.

Asian Institute of Technology 1980. *Investigation of land subsidence caused by deep well pumping in the Bangkok area.* Phase II: *Final report.* Bangkok: Division of Geotechnical and Transportation Engineering.

Berghinz, C. 1976. Venice is sinking into the sea. In *Focus on environmental geology,* R. Tank (ed.), 512-18. New York: Oxford University Press.

Biswas, B. 1973. Quaternary changes in sea level in the South China Sea. *Geol. Soc. Malaysia Bull.* **6**, 229-56.

Brand, E.W. and T. Paveenchana 1971. Deep-well pumping and subsidence in the Bangkok area. In *Fourth Asian Regional Conf. Proc. Internat. Soc. Soil Mechanics Found. Engineering,* 1-7. Bangkok: Asian Institute of Technology.

Brenner, R.P., P. Nutalaya and K. Youthong 1978. Physical and mineralogical characteristics of Bangkok subsoil from a deep borehole in Klong Luang District. In *Third Regional Conf. Geology and Mineral Resources of Southeast Asia,* P. Nutalaya (ed.). Bangkok: Asian Institute of Technology.

Cox, J.B. 1968. *A review of the engineering characteristics of the recent marine clays in southeast Asia.* Research rept. 6. Bangkok: Asian Institute of Technology.

Dawson, R.F. 1963. Land subsidence problem. *Jour. Surveying Mapping Div. Am. Soc. Civil Engineers* **89**, 1-12.

Haile, N.S. 1971. Quaternary shorelines in West Malaysia and adjacent parts of the Sunda Shelf. *Quaternaria* **15**, 333-43.

Hayes, M.O. 1964. Lognormal distribution of inner continental shelf widths and slopes. *Deep-Sea Research* **11**, 53-78.

Ishii, M., F. Kuramochi and T. Endo 1976. Recent tendencies of the land subsidence in Tokyo. *Proc. 2nd Internat. Symp. Land Subsidence,* Anaheim, Calif., 25-34.

Kelley, G.E. and S.L. Reib 1971. *Tectonic features, Gulf of Thailand Basin.* (1:5,000,000). Submitted to Department of Mineral Resources, Bangkok.

Litchfield, Whiting, Bowne and Associates 1960. *Greater Bangkok plan 2533.* Bangkok.

Metcalf and Eddy, Inc. 1978. *Report on groundwater monitoring, well construction and future programs.* Submitted to Metropolitan Water Works Authority, Bangkok.

Moh, Z.C., J.D. Nelson and E.W. Brand 1969. Strength and deformation behaviour of Bangkok clay. *Proc. 7th Internat. Conf. Soil Mechanics Found. Engineering,* Mexico City, 287-296.

Petersen, L. 1977. Mineralogical composition of the clay fraction of some soils in the Central Plain of Thailand. *Proc. 5th Southeast Asian Conf. Soil Engineering.* Bangkok: Asian Institute of Technology.

Premchitt, J. 1978. Analysis and simulation of land subsidence with special reference to Bangkok. D. Eng. thesis., Asian Institute of Technology, Bangkok.

Schmidt, K.O. undated. *Thailand (Siam) mit Stadfuhrer Bangkok,* 4th edn. Mai's Weltfuhrer no. 1 Buchenhain vor Munich: Volk und Heimat.

Wolf, D. 1978. *The five faces of Thailand – an economic geography.* London: C. Hurst.

Yamamoto, S. 1976. Recent trend of land subsidence in Japan. *Proc. 2nd Internat. Symp. Land Subsidence,* Anaheim, Calif., 25-34.

# 13

# LAND USE IN CARBONATE TERRAIN: PROBLEMS AND CASE STUDY SOLUTIONS

*Katherine A. Sheedy, Walter M. Leis,*
*Abraham Thomas, and William F. Beers*

## ABSTRACT

Suburban residential and industrial land uses in areas underlain by limestone bedrock are subject to a number of problems that are either unique to limestone areas or are accentuated by the presence of limestone. The problems are caused by the chemistry of limestone, which promotes solution of the rock along planes of weakness such as joint planes, bedding planes, lithologic contacts, and cleavage faces. Various types of problems encountered include structural instability, groundwater pollution potential, and groundwater contaminant travel.

The problems outlined above can be overcome or mitigated through planning or remedial action.

In areas that have not yet been extensively developed, regional planning for development can be instituted to prevent problems before they occur.

Where regional planning is not possible, site-specific action is necessary. This can be in the form of planning for a specific land use at a site or in the form of remedial action for existing land use.

Case study 1 describes an area-wide method for on-lot waste disposal planning; case study 2 is a methodology for site-specific planning for residential development. Remedial actions for existing problems are described in case studies 3 and 4.

## INTRODUCTION

Suburban residential and industrial land uses in areas underlain by limestone bedrock are subject to a number of problems that are either unique to limestone areas or are accentuated by the presence of limestone.

In limestone terrain, structural features such as bedding planes, fractures, joints, and lithologic contacts are enlarged by solution activity and therefore become preferred zones for water movement. The concentration of groundwater flow in these zones is unique to carbonate rocks. The result is the development of hydrogeologic regimes that are peculiar to limestone and dolostone and that create the potential for land use problems that do not occur in other lithologies. Broadly stated, these potential problems fall into two categories: those related to the structural stability of the bedrock and those related to contamination of groundwater.

Structural failure is the result of surface or near-surface collapse of bedrock which occurs most obviously in the development of sinkholes. The incidence of such events is increased by activities associated with increased land use intensity such as:

(a) Withdrawal of support in solution cavities by the installation of high-yield water supply wells and subsequent lowering of groundwater levels
(b) Concentration of surface-water infiltration at the edge of paved areas and around man-made structures
(c) Increased loading from construction of buildings or placement of fill
(d) Changed drainage patterns resulting from the installation of storm water collection systems, sedimentation and erosion control plans, and sewers

The process is thus somewhat circular, because construction and related activities that can increase the frequency of collapse events are the same activities that are most threatened by these events.

Groundwater quality in limestone aquifers is more readily threatened than in other types of bedrock aquifers, for several reasons. The presence of sinkholes and other near-surface expressions of solution activity provide an opportunity for rapid introduction of contaminants to the water table. Once the contaminants reach the water table, concentrated groundwater flow in solution channels results in rapid migration of contaminants in the system. Historically, sinkholes have been used as convenient disposal areas for agricultural, domestic, and industrial waste. This is partially attributable to the desire of the landowner to simply fill the holes, and waste material was readily available for use as fill.

The potential problems described above can be overcome or mitigated. In areas that have not yet been extensively developed, regional planning for development can be instituted to prevent problems before they occur. Briefly, such an approach would require:

(a) Categorizing areas based on the severity of solution activity.
(b) Development of building and waste disposal standards.
(c) Planning land uses that are tolerant of the potential for structural failure and that, either through design or by their nature, do not serve as potential contaminant sources.

Where regional planning is not possible, site-specific action is necessary. This can be in the form of planning for a specific land use at a site or in the form of remedial action for existing land use or existing problems. The four case histories presented are examples of:

(a) Regional planning prior to development
(b) Site-specific planning prior to development
(c) Site-specific remedial action to prevent structural failure and groundwater contamination
(d) Site-specific remedial action for existing contamination

## CASE STUDIES

### Case study 1:   areawide planning for on-lot sanitary waste disposal

The concern for proper design and construction of on-lot sanitary waste disposal systems is that the environmental conditions at the site and system design are compatible, so that sewage effluent may be disposed and dispersed within the soil with minimal long-term environmental impact. It is recognized that for all standard on-lot waste disposal systems, the soil mantle acts as a retarding medium to the migration of septic effluent into the groundwater system. In areas underlain by limestone bedrock the soil mantle is extremely variable in thickness and permeability due to solution processes. These processes, which are expressed as sinkholes, pinnacles, and disappearing streams, allow for greater-than-normal recharge rates to the water table. In addition, the water table in limestone terrain often fluctuates greatly, rising rapidly in response to precipitation infiltration, and falling just as rapidly due to flow of water through solution channels to the discharge zone. The limestone/groundwater associated problems that may be translated into on-lot waste disposal concerns are:

(a)  System malfunctions due to "short circuiting" of sewage effluent to bedrock aquifers without sufficient retardation in soil
(b)  Structural failures of the tank or drain field due to bedrock collapse

In the state of Pennsylvania, the Sewage Facilities Act (Act 537) of 1966 provided a mechanism for orderly planning and regulation of subsurface sewage systems. The County of Bucks Department of Health (BCDH) has local jurisdiction for enforcement of Act 537 in this county. Because of problems experienced in other suburban counties overlying carbonate rocks in Pennsylvania, the BCDH had enforced a prohibition on issuance of sewage system permits in carbonate bedrock since April 1, 1971. Realizing the implications of total permit prohibition, the study described was commissioned by the Bucks County Board of Commissioners to provide a basis for technical review of such permits in the carbonate belt.

A soils/geologic suitability analysis was conducted to provide the following:

(a)  Detailed map of the individual carbonate bedrock units on a 1:24 000 scale
(b)  Ranking those geologic/soils factors that are critical to subsurface sewage disposal
(c)  Evaluation and map of the suitability for sewage disposal for each of these bedrock units
(d)  Implementable approach for permit review utilizing these geologic and soils features
(e)  Translation of this approach into a "cookbook" that could be implemented by county sewage enforcement officers by modification of in-service training

To simplify and consolidate these data four soil groups and geologic features were mapped and ranked:

1. *Dolomitic/nondolomitic carbonate rock.*   This feature separates specific geologic units into broad areas of susceptibility to karst activity.

2. *Moderate- to high-yield aquifers.* Such aquifers are defined on the basis of concentration of wells having specific capacities greater than 1·5 liters/min·ft.

3. *Mottling.* Surface mottling features were mapped from 1:12 000 air photos.

4. *Sinkholes, natural depressions, and ghost lakes.* These are used as indicators of the presence of active piping features. These features were mapped from 1:12 000 air photos and USGS 7½-minute quadrangle maps.

Other factors which are specifically considered exclusionary by sewage regulations, such as floodplain location or rock outcrop belts, were omitted from the ranking.

Using a feature's presence or absence in an area, a technique called a *critical area rating system* was developed (see Table 1). In this system, which was developed for use by sewage enforcement officers (SEOs), a decision tree analysis is performed for each site. Following the format in Figure 1 accompanied by a survey plot and good-quality air photographs of the proposed site, the SEO goes through a decision process as follows:

1. A review meeting is held with the developer to discuss the water sewage needs, the development sequence, open space, and any special considerations voiced by the developer.

2. Using the initial data, the SEO determines whether the proposed site is within the carbonate belts (or an exclusionary buffer zone near contact of carbonate and non-carbonate rocks).

3. Groundwater yield potential is then determined by comparing the site location with the reported average specific capacity for each of the carbonate units that underlie the soil mantle.

4. Once these are determined the SEO scans the air photographs for specific karst features, such as mottling, sinkholes, "ghost lakes," or pinnacles.

Table 1   Site evaluation procedure for critical areas in Bucks County carbonate belts

| | Critical areas | | |
|---|---|---|---|
| Test element | I,<br>least hazardous | II,<br>moderate hazard | III,<br>most hazardous |
| Site inspection | X | X | X |
| Site mapping | (as verification) | X | X |
| Air-photograph analysis | — | X | X |
| Pit excavation | X | X | X |
| Trench excavation<br>  perpendicular to strike | — | X | X |
| Infiltration/percolation<br>  (two per acre in subdivisions) | X | X | X |
| Permeability testing for<br>  alternative systems | — | (substitute for perc testing) | |
| Subdivision review | | | |
|   Air-photograph analysis<br>  Pit excavations | X | X | X |
| Fixed-grid soil probes | — | — | X |

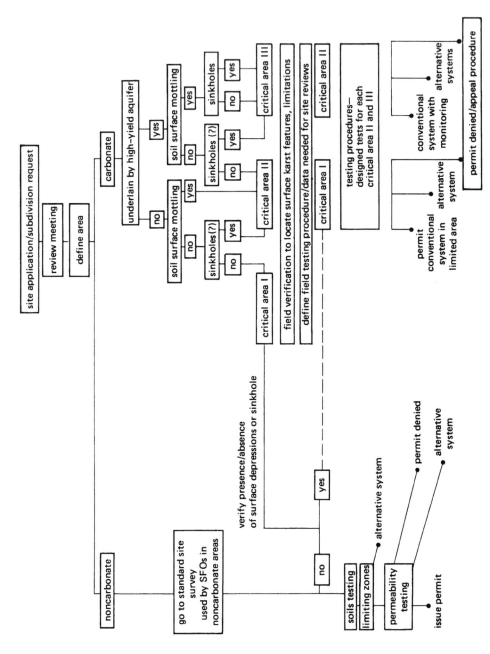

**Figure 1** Review procedure.

He or she then assigns a critical area rating to the site and field-verifies this rating for each site.

Upon completion of this initial survey the SEO reports to the developer that in the case of critical area II or III, additional data will be needed to complete the permit. These data are in the form of additional site testing to be performed by the developer's engineer.

## Case study 2: site-specific planning for rural residential development

The site in question is located in the Great Valley of Chester County, Pennsylvania, and is underlain by two carbonate rock types: Kinzers Formation (limestone) and Ledger Formation (dolomite). As seen in Figure 2, limestone is the predominant rock type. At the initiation of planning, the 127-ha site was open land that had not been farmed for at least 20 years. The proposed development was a planned rural development of 167 housing units consisting of row-style townhouses and quadraplex units. The plans specified slab foundations or 46-cm footings, depending on the unit. Immediately prior to the scheduled start of construction activities, several large sinkholes developed on the adjacent property which had recently been similarly developed. In order for the builder to continue with the project, a geologic investigation of the site to identify zones of potential structural failure was required. A phased approach was developed to maximize the data collected, provide as much opportunity as possible for the introduction of mitigative measures, and minimize interference with the construction schedule.

The initial activity conducted was a site reconnaissance prior to disturbance of the site. Air photographs were used to map surface depressions, lineation, mottled soil, and circular or linear vegetation patterns. Because of the relatively small site size, it was possible to confirm these features by walking the site. Based on these data three areas of the site showed surface expressions of solution activity. These areas were characterized by the presence of extremely shallow ($< 0.2$ m) surface depressions. They were located at the upper edge of the floodplain of a small stream and along the contact between the two lithologies (two areas) as seen in Figure 2.

Following the completion of the reconnaissance-level investigation, the locations of several buildings were shifted. The planned open space for the development was moved to the geologically active portions of the site and the buildings were relocated in areas that did not show evidence of solution activity. Shortly thereafter sinkholes began to appear in one of the active areas along the lithologic contact.

The next phase of site investigation was a drilling program which was conducted at locations of housing unit clusters. A hollow-stem auger rig was used since it was found that boring penetration into bedrock was not necessary. Areas of solution activity could be defined by the presence of:

(a) Pinnacle weathering, which was evidenced by extreme variability in the elevation of the bedrock surface. The variability that was encountered was on the order of up to 6-m changes in elevation over a distance of 3 m or less.

(b) Broken rock zones in the soil mantle or immediately above the bedrock surface.

(c) Mottled soil resulting from fluctuating groundwater levels.

(d) Solution weathering products, in this case gray silty clay, near the bedrock surface.

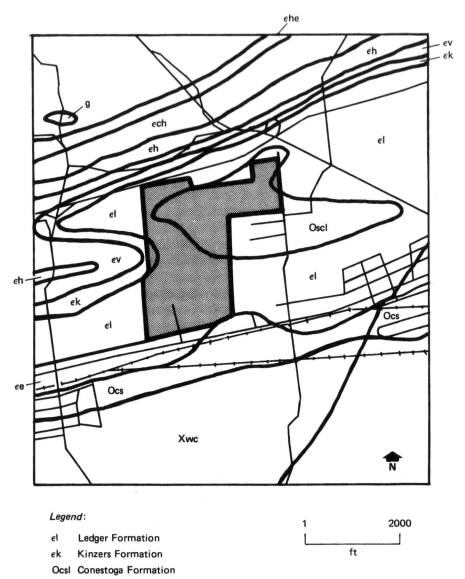

**Figure 2** Location and geology – case study 2.

*Legend*:

εl   Ledger Formation
εk   Kinzers Formation
Ocsl  Conestoga Formation

To identify areas of pinnacle weathering it is necessary to cluster borings. A spacing of three borings within 3 to 5 m was found to be suitable in this case. This need not increase the total number of borings since the spacing between clusters can be large. An example profile is shown on Figure 3.

The results of the drilling program showed that there were zones of significant solution activity in the near subsurface that did not have surface expressions. The data collected were used to make further adjustments to planned building locations. It should be noted that relocation of buildings did not decrease the number of planned housing units but did increase the total open space available on the site for passive and active recreational use.

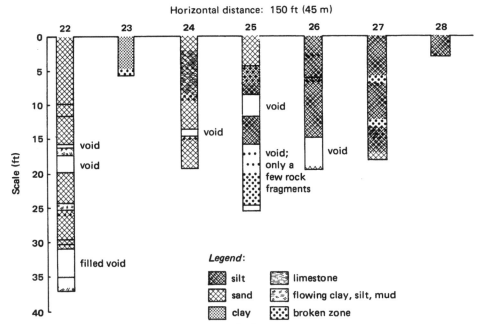

**Figure 3**  Subsurface conditions—case study 2.

The third and final phase of the investigation was inspection of excavation for sewers and footings. This was initially considered to be just a last check on site conditions but actually proved extremely useful. In three units there was evidence of pinnacle weathering in the footing excavations. This evidence included slumping of soil layers, pinnacles of weathered limestone, and bedrock exposures in the excavation walls. Because the space between pinnacles was relatively small (less than 3 m) it was possible to use steel reinforcing bars in the footings.

Construction at the site was completed in the spring of 1977. Since that time there has been no evidence of solution activity at or near any of the buildings. There has, however, been one sinkhole development in the areas designated as open space.

## Case study 3:  site stabilization

A zone of active sinkhole development was investigated at the request of a petroleum storage substation. A series of sinkholes began opening dangerously close to product storage tanks. These tanks, capable of holding 7 560 000 liters, were installed during the 1920s and because of their age, concern was voiced about their integrity. A field investigation was mounted using available outcrops and photo-interpreted lineament traces. The result of the investigation showed that the sinkholes were located along bedding planes that weathered to form pinnacles (Fig. 4). The bedding-plane pinnacles were developed as a result of differential weathering at contacts between limestone units where stylolite layers were concentrated.

The cause for the sudden appearance of these sinkholes was sought. It was quickly determined that dewatering operations in nearby quarries had lowered the water table in the area by as much as 3·3 m in the immediate area (see Fig. 4). This hydrodynamic imbalance led to subaerial collapse along the bedding planes.

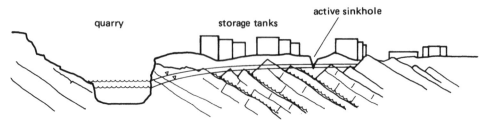

**Figure 4** Conceptual cross section.

Since time was critical from the standpoint of stability and planned use of the tanks, a grouting program using blown-in coarse angular gravel, dry cement, and netting was used to puddle the entire subsurface zone nearest the tanks. Continued surveillance by the substation engineering staff is coupled with periodic flyovers by the pipeline inspection team to monitor the appearance of further sinkzones or depressions. Additional sinkholes that have occurred outside the storage tank zone are presently being evaluated.

## Case study 4:   recovery of hydrocarbons from a limestone aquifer

This case study illustrates the extra effort required to gather and interpret data when the geomorphological characteristics of an area are marked by urbanization. The study area was an industrial facility where the boiler house was located in an abandoned, small limestone quarry. Since the facility was located within an urban area, the natural terrain features were obliterated. The initial physical evidence consisted of some rock outcrops within the quarry which indicated rock type and strike and dip of the bedding planes. Aerial photographs were not useful in identification of any significant features of karst terrain and possible contaminant migration pathways because of urbanization. Remote-sensing techniques were not suitable because of the abundance of buried drain pipes, utility lines, and so on, below the site.

Four buried oil tanks are located at the industrial facility. Two of them have a capacity of 113 400 liters each and the other two have a 56 700-liter capacity. Oil seepage was found in the steam tunnel south of the locations of the buried tanks. An oil layer had also been observed in a lined lagoon in the facility located east of the buried tanks. Figure 5 shows the location of the structures mentioned.

The regulatory agency of the region cited this industry for contaminating the groundwater with oil.

A hydrogeologic investigation of the subsurface occurrence of oil and evaluation of recovery alternatives were conducted. Preliminary investigations and examinations of the historical records indicated that:

(a) No. 6 oil was stored in the tank; first oil seepage was noticed in the main tunnel in December 1976.

(b) No oil seepage was observed in the tunnel after inspection and repair of the tanks.

(c) The creek running along the northern boundary of the property, which is at a much lower elevation than the bottom of the tanks, did not show any evidence of oil contamination.

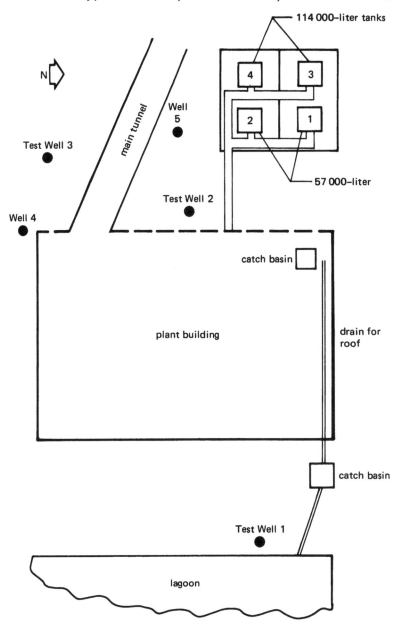

**Figure 5** Plot plan (not to scale).

(d) Oil seepage occurred only during or immediately after precipitation. Normally the oil entered into the drain located below the basement floor, which feeds into the catch basin located just outside the building. From this manhole it drained into the lagoon.

(e) No other floor drains indicated any oil contamination.

(f)  The oil that accumulated in this lagoon was being periodically removed using vacuum trucks.

(g)  The area is underlain by fractured limestone.

Based on these findings and field evaluations it was felt that the occurrence of subsurface oil in this yard was due to the past leaks from the buried tanks. This unknown quantity of No. 6 oil in the subsurface was trapped behind the tunnel wall south of the buried tanks and migrated to the storm sewer only when the water table rose. The oil ultimately discharged into the lagoon.

Upon completion of the preliminary investigation, and a meeting with the regulatory agency and the industry, a program was designed to identify and assess the magnitude of the oil seepage problem.

Initially, three on-site 15·24-cm-diameter monitoring wells were drilled and installed in the following locations:

1.  West side of the lagoon
2.  Between the buried tank and the main tunnel
3.  Immediately south of the main tunnel

Well No. 1 was drilled to a depth of 31·40 m. Bedrock, dark gray limestone with veins of calcite, was encountered at 5·34 m. Soft spots were noted around 15·24 m. Steel casing 15·24 cm in diameter was set to a depth of 7·92 m. The casing was slotted from the surface to 3·66 m for entry of oil and water into the casing. No oil contamination was observed during drilling and the water table was at a depth of about 1·52 m.

Well No. 2 was drilled to a depth of 17·98 m. Weathered limestone was encountered at 0·91 m. At 4·4 m a broken zone was noted but was dry. A solution zone was identified from 60·9 to 11·89 m below the surface.

Well No. 3 was drilled south of the tunnel to verify the theory that the tunnel wall was acting as a barrier. This well was drilled to a depth of 24·08 m. The well also encountered dark gray limestone with fractured zones from 15·54 to 18·59 m and a cavity filled with brown silt and clay from 18·59 to 20·12 m. No oil contamination was observed during drilling.

The results of the drilling confirmed that the oil that had leaked from the tanks was trapped behind the north wall of the main tunnel and that the oil level was fluctuating with the fluctuations in the water table.

Surveying of the locations and elevations of the top of the well casings was completed to determine the water table configurations. The data analysis indicated a groundwater gradient from well No. 2 to well No. 1.

Finally, monitoring of oil and water levels was conducted. Initially, an electronic oil–water interface probe was used to monitor the oil thickness. The high viscosity of No. 6 oil made it impossible to use this instrument for monitoring. A graduated surface sampler was then used to monitor oil thickness in the wells.

The monitoring of the test wells from December to the end of February 1980 indicated a 1·21- to 1·52-m drop in water table in this area as a result of a comparatively dry winter. As a result of this drop in the water table the oil–water interface dropped below the bottom of the tunnel and migrated to the south side of the tunnel. At the end of February about 2·5 cm of oil was apparent in well No. 3.

It is felt that if the facility had been located in a less developed area, the geomorphic characteristics would have been easily identified. The identification and understanding of the terrain characteristics through the use of aerial photographs, ground reconnaissance, and geologic mapping would have allowed the design of a more specific, less costly exploratory drilling program. The geomorphology of the site, if unmasked, would have given some hint of the solution cavities and fractures that acted to channelize the leaked oil.

The data described were used in evaluation of oil recovery alternatives. The oil in the wells was at a depth of about 7·01 m below the surface. The viscosity of No. 6 oil, coupled with the depth of oil in the wells, made it impossible to use surface pumps to remove the oil. A vacuum truck is now being used successfully to remove the oil from the wells. The fracture-controlled migration of oil and the viscosity of the product will not allow fast recovery of the new product from the ground. To expedite recovery, several additional recovery wells were added on-line.

## CONCLUSIONS

As we stated at the beginning of this paper, there are two approaches to solving problems posed by land development in carbonate terrain. The traditional and sometimes necessary approach has been to solve problems after they have occurred. Case study 3 is an example of a situation in which the site investigation and solution were relatively straightforward. This was due primarily to the fact that the site was in a rural area.

Since land use in the vicinity of the site was not intensive, observation of surface expressions of karst activity was possible on air photographs and in the field. Case study 4, however, is an example of the more common situation, in which the problem area is in an urban setting, with the complications and limitations of such a setting. Groundwater behavior in carbonate rocks is normally marked by complex flow systems and water levels that vary greatly in time and space. In an urban setting, the flow system is further complicated by cultural features, altered drainage patterns, and often by groundwater pumping for water supply. Field data, air photographs, and geophysical techniques are of limited value as data sources. It should be noted that in more recently developed areas, the use of historic air photographs can be extremely useful, especially in the identification of development trends that may be exacerbating karst activity.

The second approach, that of planning for development before it occurs, is obviously not always possible. Also, it is easy to view such planning as a negative factor, that is, to assume that consideration of the potential problems in carbonate terrain will automatically result in low-intensity land use. This is, in fact, not true. In both case study 1 and case study 2, more intense development than expected was possible.

# 14

# ERODE – A COMPUTER MODEL OF DRAINAGE BASIN DEVELOPMENT UNDER CHANGING BASELEVEL CONDITIONS

*N. Luanne Vanderpool*

## ABSTRACT

ERODE is a three-dimensional FORTRAN computer model which simulates the major geomorphic processes (slab failure and rockfall, rock creep, surface wash transport, gullying, and stream erosion) for slope development in a small drainage basin typical of the Colorado Plateau. ERODE is specific to the present climate and geologic conditions of the Colorado Plateau. However, it is a general-purpose model which can be used to study the consequences of changes in various conditions within the drainage basin.

ERODE has been employed to study the effects on a small drainage basin of variations in the local baselevel in the form of a relatively sudden increase or decrease. Such changes commonly occur in tributaries to major reservoirs such as Lake Powell, Utah, in response to energy and water usage demands. Results of this experiment confirm accepted ideas about baselevel: a rapid rise in baselevel results in decreased erosion, a rapid drop in baselevel results in increased erosion and stream incision. The baselevel changes result in only localized changes in the development of the basin. Long spans of time (tens of thousands of years) are required for effects to propagate upstream and upslope in the basin.

## INTRODUCTION

There is a long tradition of interest in mathematical models of slope development, starting with Fisher's classic 1866 two-dimensional model of cliff retreat. Until recently most modeling of slope evolution (Young 1963; Pollack 1969; Kirkby 1971; Ahnert 1973) has continued to consider slopes in a two-dimensional way. There are very few simulation models that actually consider slope evolution as a three-dimensional question (Ahnert 1976; Armstrong 1976; Craig 1980), yet complexities introduced by concave or convex curvature of slopes affect the landscape; there are continual changes in the directions of mass movement as the three-dimensional form of the land surface changes. Most of these models of slope evolution tend to be somewhat abstract and detached from any particular real-world situation. I have constructed a three-dimensional simulation model, ERODE, which represents the various erosional processes of slope development and stream evolution for a small drainage basin (a high plateau dissected by deep canyons) typical of the Colorado Plateau. ERODE is closely tied to this particular region and simulates specifically the processes and topographic evolution of a single drainage basin.

Mathematical models such as ERODE provide a means of experimentation in studying geological processes when other techniques are difficult or impossible. For example, there is no effective way of creating a physical scale model of topographic evolution. Although it may be possible to create the form of plateaus, slopes, and streams in such a model, it is difficult or impossible to adjust the properties and behavior of the materials so that they are consistent with scale model requirements. These problems can be partially circumvented by using mathematical simulation models which incorporate the assumed properties of the materials and their behavior through time. Direct observations of a single landscape at different points in time is the most desirable technique for yielding inferences about landform evolution. Unfortunately, because landscapes generally require long periods of time to develop, the time factor is generally prohibitive. The use of simulation models allows a landscape to be observed at different points in time as its evolution is simulated.

ERODE is a general-purpose model that can be applied to a wide variety of questions about landform evolution. It has been applied to such abstract questions as the presence or absence of dynamic equilibrium within the drainage basin, or the nature of the various interactions between the geomorphic processes in the drainage basin (Vanderpool 1979). ERODE can also be used in less theoretical and more practical directions to study the consequences of changes in certain conditions within the drainage basin. In this study ERODE is used to determine the effects of varying the local baselevel. Relatively sudden increases and decreases in local baselevel on the order of several meters within a year are produced for tributaries to major reservoirs such as Lake Powell, Utah. These changes are produced by fluctuations in the water level of the reservoirs in response to increases or decreases in energy demands or water supply and demands. Variation in the base level affects surrounding slopes in such subtle ways as a "bathtub ring" appearance – bleached and discolored areas of rock along the cliffs surrounding the reservoir above present water levels and below high-water levels. Changes in the local baselevel can have more profound effects on the development and evolution of the topography of tributaries to the reservoir. ERODE has been used to increase our understanding of these effects.

## THE MODEL

The program written in FORTRAN IV and referred to in its entirety as ERODE is the implementation and quantification of a conceptual model of slope development. This model is based upon observations and detailed surveys in an area of approximately 260 km$^2$ located about 32 km southwest of Moab, Utah, in southeastern Utah (Fig. 1). The study area is a high plateau into which canyons of more than 680 m have been carved. The resistant sandstone that forms the vertical cliffs of the area is intersected by a rectangular system of joints.

During a period of 6 years, extensive observations were made of the study area. Detailed surveys of the topography were made to illustrate the subtle variations in the landform morphology. Evidence of erosional processes was found and documented. Changes in the topographic form over this period of time were recorded to establish relative influence of the processes on the landscape. Details of these observations are presented elsewhere (Vanderpool 1979). From this came the identification of the predominant erosional processes.

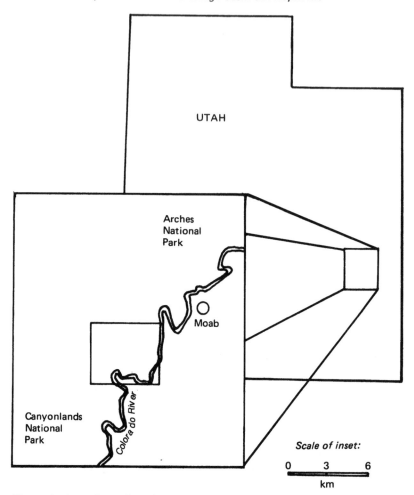

**Figure 1**  Location of study area in Utah; study area is outlined in inset.

Slope development is considered to be the response of the land surface to two groups of exopenic processes: downslope transportation and fluvial downcutting (Fig. 2). Although this model simulates a semi-arid terrain where wind would be expected to be a significant process, very little silt or sand size sediments is found unprotected by larger material (talus). As a result, wind is a quite minor process and was ignored in this model. There are two energy sources for the system defined in the simulation model. An endogenic energy source is tectonic uplift which provides the streams with gradients and provides potential energy in the form of elevation differences. The other principal energy source is exogenic in the form of solar energy which powers the hydrologic cycle and thereby drives the erosional processes overall. The processes in the system represented by the simulation model are assumed to interact with each other and with the resulting slope as shown in Figure 3. Uplift supplies potential energy for the streams and results in the regional joint pattern. Streams downcut and steepen slopes; the steepened slopes result in increased mass wasting by slab failure, rock creep of the talus, and surface wash of the colluvium. Solar energy, both via the hydrologic cycle and temperature changes, affects

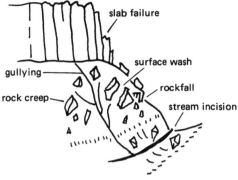

**Figure 2** Slope-forming processes on a typical simulated jointed, vertical cliff and talus-covered slope include slab failure and rockfall, rock creep, surface wash transport, gullying, and stream incision.

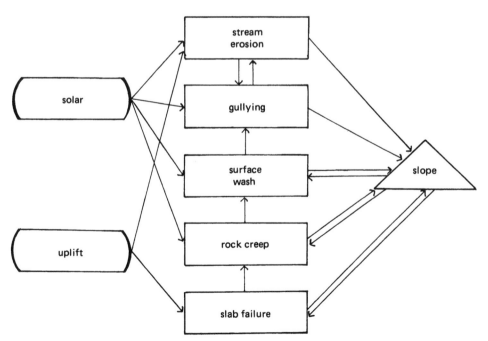

**Figure 3** System diagram showing the interrelations among external energy sources (elliptical rectangles), erosional processes (rectangles), and the slope morphology (triangle).

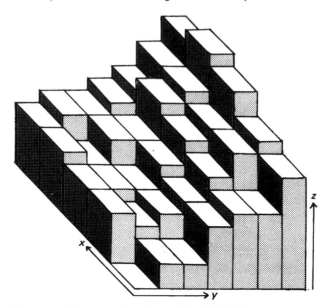

**Figure 4** System of cells used in ERODE to represent space; dimensions of cells in $x$ and $y$ directions are constant during a simulation; but in the $z$ direction, the height or elevation varies.

weathering and mass transport. The net results of the stream downcutting, mass transport, and slab failure create and modify the landscape.

In the computer program, each of the processes is represented as a separate module composed of one or more subroutines. These modules include SLABFL, which simulates slab failure; CREEP, which simulates rock creep; STRM, which simulates stream erosion; and SURFWA, which simulates downslope transport by overland flow. In addition, several other modules help maintain program organization.

In the computer program ERODE, space is represented as discrete cells (Fig. 4). In the horizontal or $x$-$y$ plane, the length and width of individual cells are constant. The vertical dimensions of a cell are defined as the average elevation of the land area represented by the cell. The horizontal dimensions of the cells represent a compromise between computer storage space and running time required (which increase rapidly as cell size decreases) and the degree of topographic detail to be represented (detail is lost as cells increase in size).

Time in the simulation model is represented within the context of the various processes. There is an internal clock in ERODE which counts time increments. But the time increment is defined indirectly by rate constants. There is a rate constant associated with each process module which gives a measure of erosion per unit time. Thus the rate constant is a quantity which calibrates the model. The rate constants are based upon rates found in the literature for similar situations within the Colorado Plateau. The rate of slab failure is based upon the rate of retreat for Threatening Rock in Chaco Canyon, New Mexico (Schumm & Chorley 1964), and of Grand Canyon, Arizona (Mortensen 1956). Rates of talus creep and surface wash are based upon measurements made by Schumm

| $F = 1{\cdot}0$ | $F = 1{\cdot}4$ | $F = 0{\cdot}5$ | $F = 1{\cdot}1$ | $F = 0{\cdot}1$ | $F = 0$ |

**Figure 5** The six possible configurations of cells along a cliff (corner cells are ignored); $F$ is geometry factor assigned to each configuration, higher values signify higher probabilities of failure. Shaded cells represent plateau top.

(1964, 1967) on slopes in western Colorado. Rates of stream erosion equal those of the Virgin River in southwestern Utah (Hamblin et al. 1975).

In order to represent a complex geomorphic system by a simulation model, many simplifying assumptions are required. A detailed discussion of these assumptions is in Vanderpool (1979). The principal assumptions follow.

(a) Slope values are calculated for each cell as the slope between the elevation of the given cell and the minimum surrounding cell.

(b) Topography represented by cells that are located along cliffs is considered to be unstable and retreats at a rate determined by a probability function that is dependent upon the state of adjacent cells (Fig. 5). The probability function values reflect the relative difficulty for slab failure by toppling determined analytically.

(c) Retreat of cliffs is accomplished by lowering the elevation of the cell to the elevation at the base of the cliff and maintaining mass balance by distributing that material as talus evenly over six neighboring cells. The distribution is monitored to prevent a buildup of material inconsistent with angle of repose.

(d) The rate at which talus is transported downslope by rock creep is assumed to be proportional to the slope angle and independent of the thickness of talus.

(e) The rate of removal of weathered bedrock and talus material by wash is a function of both slope angle and distance from a drainage divide.

(f) The slope is assumed to be transport limited (Carson & Kirkby 1972); there is more weathered material available to be removed than there is ability to transport the material. As a result, the volume of available weathered material is not calculated.

(g) Streams are located in the simulated space as they occur in the natural landscape being modeled. For a cell that contains a stream, stream erosion is assumed to remove bedrock as a function of stream gradient and discharge (Howard 1970).

Before ERODE was used for any experiments, it was tested to ensure the consistency of the assumptions concerning the relative rates of processes and modes of interaction of the processes and to double check the reasonableness of the rate constants. Using a simple initial topography, the model was allowed to run for 200 time increments (about 40 000 years). The resulting simulated landscape was compared with the topography in the study area. After 50 increments the model produced a topography similar to the topography of the study area and maintained that topography throughout the simulation. The resulting rate of erosion, approximately 0·1 mm/yr, compares quite favorably to measured average denudation rates in the Colorado River Basin, 0·16 mm/yr, since the basin-wide rate also includes the more rapid denudation of some mountainous terrain within the basin (Smith et al. 1960).

## RESULTS OF BASELEVEL CHANGES

To allow the study of the effects of various changes in baselevel, the process module STRM contains options which simulate four alternative baselevel possibilities:

1. No baselevel control; elevation at the downstream mouth of the drainage basin is free to change in response to erosion and deposition within the simulated basin.
2. The baselevel is constant; the elevation at the downstream mouth of the drainage basin is held constant.
3. The baselevel is rapidly dropped; at a particular time increment, elevation at the downstream end of the drainage basin is dropped to the new baselevel.
4. The baselevel is rapidly raised; after a particular time increment, the mouth of the drainage basin is flooded to the elevation of the new baselevel and no elevations lower than the new baselevel are allowed at the downstream end. During a single simulation, only one of these four options may be selected.

By comparing the results of simulation runs of ERODE using each of these options, it is possible to examine the effects of several different baselevel conditions on the topographic form through time. I simulated changes in the landscape of a single small

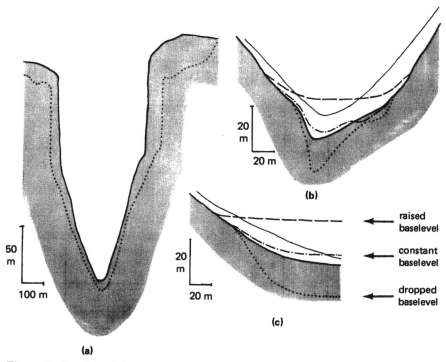

**Figure 6** Results of the baselevel experiments: (a) shows entire valley cross section 60 m upstream from mouth at beginning (solid) and end (dotted) of experiment 1; (b) details of the valley cross section; (c) stream profile at mouth of valley with baselevels indicated. In (b) and (c) light solid line is original topography, heavy solid line indicates results of experiment 1; dotted and dashed, experiment 2; dotted, experiment 3; and dashed, experiment 4.

**Table 1** Results of the baselevel experiments

| Experiment | Baselevel effect | Overall average elevation change (m) |
|---|---|---|
| 1 | No control | 3·665 |
| 2 | Constant baselevel | 3·645 |
| 3 | Baselevel dropped | 3·754 |
| 4 | Baselevel raised | 3·509 |

drainage basin (approximately 1 km² in area) located within the study area of Figure 1. In each case, the beginning landscape is the present topography. The model ran for 200 units of time iteration, which represents about 40 000 years. In all cases the general form and morphology of the landscapes are similar; valleys widen and deepen slightly during the simulation. Only in detail, in the areas downstream near the valley mouth, do the simulation results differ. Experiment 1 (no outside baselevel control) serves as the normal or experimental control for this investigation. To identify changes due to base-level controls, results of the other runs are compared with the results of experiment 1 (Fig. 6). In general, baselevel controls have limited effects and move slowly through the system. After the full 200 iterations, no changes caused by baselevel difference occur farther than 100 to 120 m from the mouth of the drainage basin. For easy comparisons of rates of erosion between experiments, overall volume change in cells from erosion is tallied for all cells throughout the simulated drainage basin. This net volume change divided by total area yields an average elevation change for each experiment; a single number, which although not the traditional method for measuring erosion rates, is convenient for comparing the results of these experiments (Table 1).

In experiment 2 the baselevel is held constant. The overall average elevation change, 3·645 m, is lightly less than in the control (3·665 m), since the rate of degradation was greatly reduced around the downstream valley mouth. But the effect is local; after the entire 200 increments all cells located more than 120 m from the mouth of the valley remain unchanged. In experiment 3 the baselevel is dropped. A knickpoint forms at the mouth of the valley and slowly advances up the valley. Seventy-five time increments after the baselevel drop, the knickpoint had advanced 40 m. After the entire 200 time increments the knickpoint had advanced a total of 80 m. The overall amount of degradation (3·754 m) is much greater. This increased erosion is largely due to stream incision; no measurable valley widening occurs. Experiment 4, in which the baselevel is suddenly raised and the mouth of the valley flooded, results in a sharp decrease in the overall erosion rate (3·509 m) compared to the control experiment. This decreased erosion is limited once again to less than 120 m from the valley's mouth.

## PRACTICAL IMPLICATIONS

This study has investigated the influence of variations in local baselevel conditions on erosion in drainage basins. It was intended to approximate the conditions found in drainage basins that are tributaries to major reservoirs such as Lake Powell, Utah, or Lake Mead, Arizona. Changes in the water level of the reservoir from initial filling, draining, or replenishment are equivalent to raising or lowering the local baselevel for the tributary basins. These simulation experiments show that a sudden drop in baselevel, such as would occur when a reservoir is drained for maintenance or when high water demand downstream consumes much water from the reservoir, results in decreased degradation within the tributary valley. A rise in local baselevel, which would occur when the reservoir is initially filled or is replenished, results in decreased erosion for the tributary basin. When baselevel rises, some previously subaerial land is covered by water, which results in a loss of usable surface land. Although less erosion takes place in the drainage basin, which would suggest increased stability for the remaining exposed land, the newly submerged land will tend to be subjected to increased weathering. This is particularly true for the many rocks of the Colorado Plateau with high calcium carbonate content whose erosional resistance is due to the dryness of the climate. Later subaerial exposure of these rocks will find them with enhanced susceptibility to mechanical weathering and erosion.

When baselevel drops, there will be a gain in subaerial land, some in the form of small beaches composed of sediment derived from the basin and deposited in the reservoir. But this is balanced by increased erosion of all the land. The newly exposed land will be unstable and possibly less resistant to erosion than will surrounding land, owing to the augmented weathering during submergence. The stream will begin to incise; this incision can affect land upstream from the region directly affected by the baselevel drop, the mouth. The incision will reduce the amount of available stable and usable land.

Overall, the effect of varying local baselevel is a reduction in usable subaerial land. If baselevels fluctuate up and down, this process would be intensified and greater land loss would occur. As the erosion is increased in the tributary basins due to the water-level changes, increased amounts of sediment will be deposited in the reservoir, limiting the usefulness of the reservoir itself. This would suggest that to protect both the life of the reservoir and the subaerial land around the reservoir, once the reservoir is filled the water level should be held constant.

## CONCLUSIONS

Results from the baselevel investigations are not surprising; the effects of all changes in baselevel conditions result in only localized changes in the landscape of the drainage basin. This is reasonable and expected since baselevel effects are transmitted through the system by the stream network (Schumm 1977). This study shows that long spans of time (on the order of tens of thousands of years) are required for effects to propagate upstream and upslope throughout the basin.

The results of the study produced no new or controversial insights about base level. They contradicted no accepted geomorphic concepts about baselevel. On the contrary, the use of ERODE supported, by a new line of evidence, ideas about baselevel which are not usually subject to experimentation. The simulation allowed a practical problem, the implications of water-level changes in reservoirs on the erosional development of the surrounding region, to be studied. The results suggest a possible management policy – water levels should be held constant to minimize sedimentation of the reservoir. The strength of simulation models such as ERODE rests in their flexibility, as media to be used in various experiments involving landform change through time, such as during this study.

Further investigations along the same lines as those used in this study might simulate much longer periods of time in order to examine the propagation upslope and upstream of effects due to these baselevel changes. Alternatively, the simulation might incorporate baselevel fluctuations up and down in a single run. However, even more desirable would be a simulation with significantly smaller cells over a shorter period of time. This would allow the recognition of changes in the downstream topography in greater detail than in this more general study.

## REFERENCES

Ahnert, F. 1973. COSLOP-2: a comprehensive model for simulating slope profile development. *Geocom. Programs* **8**, 1-24.

Ahnert, F. 1976. Brief description of a comprehensive three dimensional process-response model of landform development. In *Zeitschr. Geomorphologie* N.F. Suppl. **25**, 29-49.

Armstrong, 1976.

Carson, M.A. and M.J. Kirkby 1972. *Hillslope form and process.* Cambridge: Cambridge University Press.

Craig, R.G. 1980. A computer program for the simulation of landform erosion. *Computers Geosci.* **6**, 111-42.

Fisher, O. 1866. On the disintegration of a chalk cliff. *Geol. Mag.* **3**, 354-6.

Hamblin, W.K., P.E. Damon and M. Shafiqullah 1975. Rates of erosion in the Virgin River drainage basin in southern Utah and northern Arizona. Geol. Soc. Ann. Mtg., Salt Lake City.

Howard, A. 1970. A study in form and process in desert landforms. Ph.D. dissertation, John Hopkins Univ.

Kirkby, M.J. 1971. Hillslope process-response models based on the continuity equation. In *Slopes, form and process*, D. Brunsden (ed.), 15-30. Inst. British Geog. Spec. Pub. 3.

Mortensen, H. 1956. Uber Wandverwitterung and Hangabtragung in semiariden and volariden Gebieten. International Geographical Union: First report on the study of slopes, 96-104.

Pollack, H. 1969. A numerical model of the Grand Canyon. In *Four Corners Geol. Soc. Guidebook to 5th Field Conf.* 61-62.

Schumm, S.A. 1964. Seasonal variations of erosion rates and processes on hillslopes in western Colorado. In *Zeitschr. Geomorphologie* N.F. Suppl. **5**, 215-38.

Schumm, S.A. 1967. Rates of surficial rock creep on hillslopes in western Colorado. *Science* **155**, 560-1.

Schumm, S.A. 1977. *The fluvial system.* New York: Wiley-Interscience.

Schumm, S.A. and R.J. Chorley 1964. The fall of Threatening Rock. *Am. Jour. Sci.* **262**, 1041-54.

Smith, W.O., C.P. Vetter, G.B. Cummings et al. 1960. Comprehensive survey of sedimentation in Lake Mead, 1948-49. U.S. Geol. Survey Prof. Paper 295.

Vanderpool, N.L. 1979. Mathematical model of landform development in the Canyonlands region of southeastern Utah. Ph.D. dissertation, Stanford Univ.

Young, A. 1963. Deductive models of slope evolution. *Nachr. Akad. Wiss. Goettingen, Math. Phys. Kl.* **2**, 45-66.

# 15

# MORPHOLOGIC AND MORPHOMETRIC RESPONSE TO CHANNELIZATION: THE CASE HISTORY OF BIG PINE CREEK DITCH, BENTON COUNTY, INDIANA

*Robert S. Barnard and Wilton N. Melhorn*

## ABSTRACT

Big Pine Creek in northwestern Indiana drains an extensive area of fertile prairie soils developed on Wisconsinan age till and alluvial deposits. The creek was channelized in 1932 to improve drainage; original channel geometry, slope, and sinuosity can be approximately restored from the engineering plans and design criteria. No organized or consistent maintenance has been performed on the stream, and by using time-sequential aerial photographs for the years 1938, 1963, and 1971, a topographic map prepared in 1962, and a field survey of selected reaches in 1975 it is possible to monitor progressive changes over a recovery interval of about 40 years. Principal adjustments include (a) an increase in sinuosity and initiation of new point bars and meanders accompanying lateral migration of the channel, (b) reestablishment of pool-riffle sequences through scour and fill in parts of the channelized reach, (c) bank erosion and slumping, (d) a decrease in gradient and other profile adjustments, and (e) reduction of channel drainage capacity with an increase in flood frequency and magnitude. These adjustments, however, are neither uniform nor consistent over the entire channel, probably because of interim maintenance undertaken by individual farmers along discrete segments.

Attempts to determine the principal cause of channel deterioration include an investigation of variations in bank material, distribution of spoil materials, and threshold values related to variations in sinuosity, slope, and stream power. The relationship between sinuosity and stream power seems most significant, and a stability envelope is developed for the natural, prechannelized stream. Assessment of this relationship from data available from aerial photographs and the field survey suggests that though instability continues there is a long-term adjustment trend toward the prechannelized state. This trend can also be used to predict recovery rates and cumulative recovery with time.

It is suggested that if stream power to sinuosity and other relationships are identified, quantified, and properly applied, engineering design error may be greatly minimized, and if geomorphic thresholds are not exceeded, a stable, effective drainage project may be constructed without significant loss of biotic, vegetational, or aesthetic qualities of the stream.

# INTRODUCTION

## Geographic setting

Benton County, in northwestern Indiana about 65 mi (105 km) south of Lake Michigan, is a corn and soybean farmland district which reputedly has the highest dollar-value agricultural productivity of any area of equal size in the world. The county lies in the headwater region of several major watersheds; drainage east, west, and north flows eventually to the Iroquois and Tippecanoe Rivers and southward to the Wabash River. The major Benton County stream, Big Pine Creek, heads in an east-west-trending kame-moraine divide and flows south across an almost flat expanse of Wisconsinan age ground moraine and outwash and Recent alluvial and aeolian deposits (Fig. 1).

## Prechannelization history

Settlement commenced along Big Pine in the 1830s, and by 1845 agricultural encroachment reached malaria-infested, higher-level blue-stem prairie grasslands where soils were wet but good for farming. Small open ditches and crude tile drains were installed to facilitate drainage by about 1875. However, as agricultural development intensified and additional prairie sod was removed for cultivation, flooding became common and increasingly persistent by the early 1900s, and each year valuable topsoil was lost to

**Figure 1** Location map of Benton County, Indiana, and Big Pine Creek. Channelized reach and knickpoint are indicated. End moraines shown schematically; blank areas are ground moraine.

flood flow. In 1930 local farmers formed an independent consortium to channelize a 10·1-mi (16·25-km)-long reach encompassing all of the west branch of Big Pine, from the headwaters to the juncture with a previously dredged reach of the east (main) branch of the creek.

Only sparse information is available on natural stream regimen and channel parameters prior to channelization. However, from a planimetric map made in 1878 and because spoil from the dredging project was poorly distributed and not consistently used to fill the preexisting channel, original channel geometry and sinuosity can be fairly accurately restored by taking measurements from meander scars and old channel patterns as seen on USDA 1:21 120 scale aerial photographs taken in 1938, six years after project completion. The preconstruction channel longitudinal profile can be estimated from elevations obtained from current USGS 1:24 000 scale topographic maps. Although cross-sectional data could not be reliably approximated as unfilled portions of the original channel have since infilled by slope wash or been modified by cultivation, it appears that Big Pine Creek was a small, extensively meandering stream (sinuosity = 1·42) with a relatively wide floodplain. Total fall was about 65 ft (19·5 m) through the channelized reach, an average slope of 6·3 ft/mi (0·11 m/km). Meander wavelength varied from 100 to 450 ft (30 to 135 m) with a mean value of 300 ft (15 m); meander amplitudes ranged from 75 to 300 ft (22·5 to 90 m) with a mean of 150 ft (45 m).

## Engineering construction

The privately funded project was completed in June 1932 and fortunately the original engineering plans and design criteria still exist (Benton County Court Records 1930–32). The channel was designed as a box-shaped trapezoid with base widths varying from 4 ft (1·2 m) at the head to 12 ft (3·6 m) at the mouth, and side slopes of 1·5 to 1·0. At project completion (the cost was only $15,000, or 60% of the budgeted amount!) the channel drained 15·88 mi² (25·57 km²), had been shortened 2.9 mi (4·7 km) (a decrease of 30%), sinuosity was reduced from 1·42 to near 1·00, and the average channel slope was increased from 6·3 ft/mi (3 m/km) to 8·5 ft/mi (4 m/km). The construction cost included removal of 279 yd³ (254 m³) of limestone bedrock along a 200-ft (60-m) section of channel near the midpoint of the channelized reach. It is not known whether the stream flowed on this bedrock "high" (Fig. 1) prior to channelization, but the rock ledge plays a role in post-channelization profile adjustment and has formed a knickpoint in recent years.

For purposes of our research on the (now) Big Pine Creek Ditch, it was assumed that the project was conducted in accord with plans and specifications, as on completion the ditch was inspected and approved by the county surveyor and engineer-in-charge. However, design plans specified that spoil should be evenly distributed on both banks, leaving a 5-ft (1·5-m)-wide berm between the edge of ditch banks and the toe of spoil banks, whereas our field examination showed that spoils were not evenly distributed or used to fill portions of the preexisting channel. Other minor deviations from the original plans thus probably also occurred.

## CALCULATIONS FROM DESIGN CRITERIA

Although no discharge measurements or flood frequency curves were included with the engineering design (nor are these data available for the stream today), several methods for estimating discharge were tried, ranging from simple empirical formulas to more complex regression-type equations. These procedures yielded widely varying results, but the regression method developed by Davis (1974) for estimating discharge for varying recurrence intervals in small watersheds in Indiana seems the best. Discharges were then estimated for six points on the stream where cross-sectional area was increased according to engineering plans. Combining the Davis regression equation with the area–velocity technique and Manning's equation [with $n = 0.03$, as recommended by Gregory and Walling (1973) for clean, straight streams with no deep pools], it was found that the design channel would accommodate a 25-year flood over its entire length. However, calculations showed also that in reaches where the channel base width was 6 and 10 ft (1·8 and 3 m), respectively, a 50-year flood event would exceed channel capacity. Otherwise the design parameters for width, cross-sectional area, and slope were appropriate for a drainage basin of this size.

No organized or consistent maintenance measures have been performed since 1932. Some reaches have been maintained by individual farmers, whereas other sections have steadily deteriorated in both plan and profile, there has been considerable bank slumping and trees, brush, and weeds again encroach on the channel.

## MORPHOLOGIC AND MORPHOMETRIC CHANGES

### Channel adjustments

An historical evaluation of channel deterioration in terms of magnitude and rates of change in pool-riffle sequences, sinuosity, gradient, bank erosion, and redevelopment of natural meanders is possible using sequential aerial photography obtained in 1938 (scale 1:21 120) and in 1963 and 1971 (scale 1:15 840). Some critical reaches were field surveyed in 1975, and included measurement of channel cross sections and plane table mapping on a 2-ft (0·6-m) contour interval.

Yearke (1971) has noted that most major adjustments should occur immediately after construction as a channel seeks to restore a natural balance. No improvements were undertaken on the main channel between 1932 and 1938, so the local base level of Big Pine Creek Ditch remained essentially constant, yet 1938 aerial photos show that considerable meander development occurred just upstream from the ditch outlet and in another reach farther upstream that had been constructed with relatively high gradients (0·0014 and 0·0027 slopes, respectively), whereas no change is noted in an intervening reach which had a lower slope (0·0010). It could be speculated that a threshold effect occurs between the slope values of 0·0010 and 0·0014, but the data are sparse and the impacts of interim, locally remedial channel maintenance cannot be quantitatively assessed.

In another reach, by 1938 the channel already resembled the prechannelized pattern; however, we believe that construction at this point had insufficiently filled the original channel so that it was reoccupied during periods of high flow. The problem area does not reappear on 1963 photography, so the situation must have been rectified and the channel repaired before additional deterioration took place. Unfortunately, no data are available on profile adjustments immediately following channelization, but probably some scour and fill occurred during the initial readjustment period. Profile adjustments at this time would probably also initiate lateral migration of the channel and promote incipient meandering.

**Figure 2**  Upstream view just above mouth of Big Pine Creek Ditch, showing bank erosion in unevenly distributed spoil material, amplitude of new meanders, and formation of point bars and incipient pool-riffle sequences.

By 1963 considerable lateral as well as profile adjustment is apparent. Every reach that was deteriorating in 1938 continued to degenerate, in the sense that fluvial processes operated to progressively alter or destroy the alignment, width, or depth of the engineered channel. Meander development in the outlet reach had increased in frequency and amplitude (Fig. 2) and seems to have propagated farther upstream into the low-slope

**Figure 3**  Upstream view of meandering, low-water channel in the headwater reach of channelized Big Pine Creek. Bank deterioration is minor but the channel is characterized by sinuous reaches broken by a few straight segments.

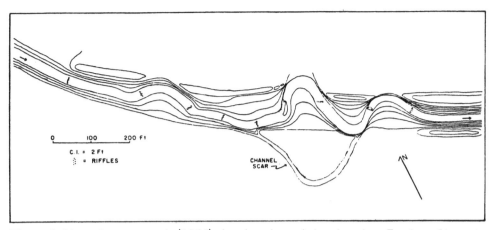

**Figure 4** Map of survey reach (1975) showing channel deterioration. Erosion of bounding spoil banks and formation of meanders, point bars, irregularly spaced riffles, and a midchannel bar are clearly evident.

sector that was still essentially straight in 1938. The high gradient reach farther upstream now has considerable sinuosity (Fig. 3) but is broken by a few straight segments. The coincidence of property lines at the breaks between straight and meandering segments suggests that individual landowners had performed some interim or more frequent maintenance on the channel across their property. An area of bedrock forms a knickpoint at about the midpoint of the channelized stream. Comparison of longitudinal profiles in this

**Figure 5** Downstream view in survey reach showing bank erosion and vegetated point bar. The truncated line of the spoil bank appears as the ridge just to the right of the tree.

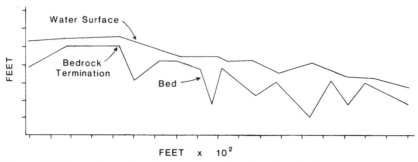

**Figure 6** Channel profile of survey reach indicating development of an advanced stage of pool-riffle development (vertical exaggeration is 80X).

channel sector for the years 1932 and 1962 yields results that closely approximate those described by Holland and Pickup (1976) for profile adjustments on knickpoints in stratified sediments in a laboratory flume: (a) an aggraded reach just upstream from the knickpoint termination, but passing into an oversteepened reach just above the knickpoint face; (b) a steepened knickpoint face; and (c) an incising reach downstream.

In summary, in the 40 years since channelization, readjustments toward a more natural equilibrium condition apparently occurred at varying rates and at different times. That is, some reaches show little if any deterioration of the constructed channel, whereas others show extensive meander development and channel migration and have reached advanced stages of pool-riffle development. Several reaches show excessive bank erosion

**Figure 7** Downstream view in survey reach showing bank erosion, point bar deposition, and formation of riffles.

and lateral migration of the channel point bars and are accompanied by varying degrees of entrenchment. Sinuosity in some reaches now approximates 1·2.

In 1975 we surveyed a channel reach that is experiencing excessive bank erosion (Fig. 4), where clearly there is bank cutting occurring on the outside of meanders and point bar deposition on the inside of bends (Fig. 5). Figure 6, a profile of the channel bottom through this reach, indicates that an advanced stage of pool-riffle development has accompanied meander adjustments (Fig. 7).

## Flooding

There is no station precipitation record nearer than Lafayette, 50 mi (80 km) distant, where the climatic record shows no unusual rainfall anomalies or long-term trend between 1930 and 1975. Agricultural practices in the basin have changed somewhat over the years in that there are now fewer fence line buffer zones of vegetation, but any tendency toward an increased sediment loss by their removal is probably compensated for by improved tillage techniques and the trend toward decreasing erosive losses by armoring fingertip tributary channels with a grass mat. However, local residents everywhere along the ditch again complain about the magnitude and frequency of recent flooding. Assuming that discharge has not appreciably increased in 40 years, deterioration of the channel is seen as a probable cause of these conditions. Cross sections accordingly were made at several points in a 7 mi (11-km)-long study reach; these show that locally there has been only minor deterioration, whereas much greater amounts occur elsewhere (Fig. 8). Figure 8a, taken near the headwaters, shows only slight changes since 1932. Conversely, Figure 8b and c document considerable bank slumping, collapse, entrenchment, and deposition on the inside of meander bends. These cross sections were made in both straight and meandering reaches to show different types of degeneration, slow slumping and creep of banks in straight reaches, and lateral migration of channel and meanders causing undercutting, bank collapse, and point bar deposition on the inside of meander bends. In the survey reach, bank deterioration and minor downcutting definitely has increased the channel cross-sectional area below the knickpoint. Concurrently stream gradients have greatly decreased and trees and brush encroach on the channel banks. Applying Manning's equation to the present cross-sectional area and slope, and using an $n$ value of 0·07, maximum velocity has been reduced from 5·2 ft/s (1·6 m/s) in 1932 to 2·6 ft/s (0·78 m/s) in 1975. Despite the increase in cross section, the channel has lost as much as 60% of its capacity in the reach. Comparing the present maximum capacity of 1100 ft³/s (330 m³/s) [425 ft² (127·5 m²/s) cross-sectional area × 2·6 ft/s (0·78 m/s)] to figures obtained for the 1932 design capacity shows that this reach, which formerly could accommodate a flood magnitude in excess of a 50-year event, can now scarcely accommodate a 25-year flood recurrence. Even if only approximate, these calculations demonstrate the effects of channel deterioration and increased bank vegetation on stream capacity.

## Pools and riffles

Study of channelized stream reversion logically leads to an investigation of the development, spacing, and configuration of pools and riffles. Obviously, the modified channel had no pool-riffle sequences and probably very little of any type of bedform variation. However, varying flow conditions and sediment loading will soon allow scour and fill processes to operate in different sections of the stream bed and will form asymmetric shoals on alternating sides of the channel. Present conditions on Big Pine Creek tend to

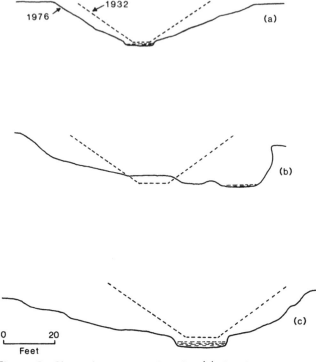

**Figure 8** Channel cross section in (a) headwaters reach shown in Figure 3, (b) survey reach shown in Figures 4 through 7, and (c) 0·5 mi (0·8 km) upstream from outlet, in reach shown in Figure 2. Dashed lines indicate approximate configuration immediately after construction; solid lines are the configuration in 1976. The water surface level is approximate.

conform with the initial stages of the five-stage model of the evolution of channe bedforms as proposed by Keller (1972). Headwater reaches that have low flow volume and sediment load lack pools and riffles, whereas farther downstream asymmetric shoa development becomes predominant, followed by reaches having pool-riffle sequence with varied spacing (commonly three to five channel widths). Most reaches can be classe as fitting within the first four stages of the Keller model; mature, well-developed, stabl pool-riffle sequences spaced at five to seven times channel width (stage 5 of the Kelle model) will probably never occur on a channelized stream, as this requires a return to near-equilibrium state with close to the original sinuosity. The most advanced pool-riffl development on Big Pine tends to coincide with reaches of greatest meander developmen and seems to fit into the Keller stage 4 in having pools in each bend, riffles at the inflec tion between meanders, and spacing of insufficient length to indicate channel stability Even so, the return to a more natural distribution of bed forms, such as pools and riffle is likely to be fairly rapid as compared to biological recovery in the creek. If a stabl bedform state is ever reached again, biotic stability will necessarily take longer than mo phologic recovery, as suitable habitats are dependent on achievement of a nearly stabl channel.

# GEOMORPHIC THRESHOLDS

It is important to study the problem of geomorphic thresholds in channelized streams, as such streams present a relatively unusual setting. As in a laboratory stream (flume), channel slopes in modified streams are set on specific gradients rather than allowed to evolve naturally. If such gradients are not held constant but vary among different reaches, some reaches may develop pool-riffle sequences or show meander development, whereas others may not.

This suggests that channel morphology may not respond gradually to alteration of parameters such as slope, but may tend to contain thresholds, or critical values, below which response will be negligible and above which the response may be quite rapid. Schumm and Khan (1972) demonstrated such threshold values in the laboratory by varying slope and noting the sinuosity of the adjusted stream. Results indicated that thresholds occur between straight (less than 0·002 slope) and meandering channels (0·002 to 0·015 slope), and between meandering and braided channels (more than 0·015 slope). Similarly, Edgar (1973), through flume studies and measurement of natural conditions on thirty-seven stream reaches in Alberta, attempted to show that variations in stream power (slope X discharge X specific weight of water) were the cause of changes in sinuosity, and that sinuosity varied with slope and discharge, not with slope alone.

Other parameters, such as momentum, bank material type, and sediment load, have been varied experimentally to yield similar results. Assuredly a determination of threshold values could be of great benefit in channelization and river-control projects. If the threshold values can be determined prior to construction, engineers could estimate how much straightening or slope modification a given stream could tolerate without exceeding existing thresholds. Thus planning and design could be carried out with greater confidence that destructive aftereffects would be minimal.

We attempted to define the relationship of such threshold parameters to phenomena observed on Big Pine Creek Ditch. Investigation was made of the relationship of slope and stream power to sinuosity to see if any meaningful correlation existed. Sinuosities were measured for the entire study reach and were divided into different segments, that is, into separate 1-mi (1·6-km)-long reaches, and also into reaches with a constant 9 ft (1·5 m) of fall. Slopes were measured from the channel profiles of the prechannelized stream and the channel of the 1932 and 1962 streams.

It is also generally accepted that channel morphology, and thus sinuosity, is formed or determined by a bank-full discharge stage. As no cross-sectional or slope data are available for the prechannelized stream, estimates must again be used and, applying Davis's (1974) method, the 2-year flood is determined as the closest approximation to bank-full stage. Estimates were made for numerous drainage areas or subareas within the 16-m$^2$ (26-km$^2$) basin.

The Schumm–Khan relationship of slope to sinuosity was tested and found inadequate for explaining observed conditions on Big Pine Creek. However, a plot of stream power to sinuosity relationship (Fig. 9) for the prechannelized stream produces some interesting results. It is apparent from this figure that slope and thus stream power on the Big Pine do not vary sufficiently to produce either straight (sinuosity = 1·00) or braided reaches, but continuation of the curve, or trend, to lower slopes would indicate that a threshold condition does exist. However, the validity of this plot can be questioned, as different curves could be drawn for the data provided.

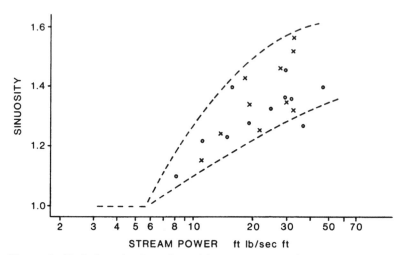

**Figure 9** Variations in sinuosity with stream power for prechannelized conditions on Big Pine Creek.

If Big Pine Creek was in a near-equilibrium state prior to channelization, a curve or "envelope" developed for that time period represents the stream in its most stable condition. The authors believe that any deviations from this envelope represent instability and thus indicate that readjustment is forthcoming. Channelization produces just such deviations to an extreme degree by reducing sinuosity to 1·0 and concurrently increasing slope. In theory, any movement within the envelope will maintain the stability of the natural state. In channelization, if sinuosity is reduced, then slope, and thus stream power, must also be reduced. This can be accomplished in the conventional channelization process only by using drop or weir structures.

It is anticipated that for any point in time in the readjustment process, slope or stream power curves could be constructed to represent the stream at that time. It is also believed that these curves will tend to migrate upward from a sinuosity of 1·0 in the direction of the prechannelization stable curve already determined. Such curves were developed for the stream as it existed in 1938, 1963, and 1971, using slopes obtained from the 1932 design and 1962 topographic map profiles.

The stream power to sinuosity curves appear in Figure 10, with different years represented by different symbols. Although the process of curve fitting through these points is more difficult than for the prechannelized stream, definite trends are indicated. Some very low points can be eliminated in the fitting process, as they were probably caused by interim maintenance measures undertaken by individual farmers.

More information can be gained from these curves than has previously been suggested. A definite migratory trend is evident in curves representing the years 1938, 1963, and 1971, with progressive adjustment in the direction of the prechannelized state. A rate of recovery can be estimated from these curves; however, caution must be exercised, as this rate may tend to fluctuate with time and may also differ among various reaches of the stream. In other words, recovery rates may be increasing in some reaches but simul-

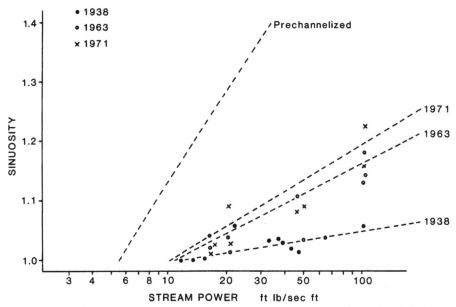

**Figure 10** Variations of sinuosity with stream power on Big Pine Creek Ditch in the years 1938, 1963, and 1971.

taneously decreasing in others, and general trends in recovery may experience fluctuations spread over the entire 7-mi (11·25-km) reach.

Although meander development in some reaches currently appears to be experiencing accelerated recovery, rates over the entire stream seem to be decelerating. Recovery rate fluctuations may also partly explain the point spread that appears on the stream power to sinuosity plots in Figure 10.

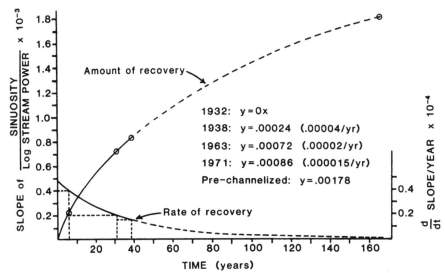

**Figure 11** Estimated amounts and rates of recovery with time, from channelization in 1932 until total recovery after 165 years.

For Figure 10 lines were fitted by visual inspection to each set of points representing each time period, and the slope for each line was calculated. These slopes represent a curve of stream power related to sinuosity (as plotted on semilog paper, they approximate straight lines) for the 40-year period. These slopes were then plotted against time to produce the upper curve shown in Figure 11. Extension of this curve to the slope of the prechannelized stream power vs. sinuosity line thus gives some estimate about the time required for the stream to regain equilibrium conditions. The projection shows that Big Pine Creek Ditch, if allowed to continue natural adjustment with no remedial maintenance, by the year 2080, after 165 years of recovery, would again approximate the prechannelized quasi-equilibrium state.

The lower curve in Figure 11 shows the rate of change in recovery, and the area below the curve indicates cumulative recovery. Extension of this trend indicates, as predicted, that recovery rates were high initially and have since been decreasing. This does not, however, discount the fact that some reaches are presently experiencing accelerated erosion and thus a rapid increase in sinuosity.

## VALIDITY OF RESEARCH METHODOLOGY

There are several steps in the methodology that depend totally on estimation. Although these estimates seem conservative and accurate, there is room for error. The error-prone areas are: (a) misinterpretation at any investigation phase of the 1938, 1963, or 1971 photography; (b) deviation between actual construction and the approved plans and specifications for the channelization project; (c) error in discharge estimates and use of a 2-year flood event as equivalent to bank-full stage; (d) misuse of topographic maps, as contours were used to draw profiles of the prechannelized stream and the stream in 1963; (e) mistakes in velocity calculations because of incorrect estimation of the appropriate Manning $n$ value; and (f) fitting stream power curves "by eye."

## APPLICATIONS TO ENGINEERING DESIGN

Although much of the foregoing discussion of thresholds and recovery rates is tenuous, we believe that further testing and application of such concepts could lead to better design of stable channels. The envelope for the stream power-to-sinuosity relationship seems to represent the stable condition or that of a quasi-equilibrium state. Points lying within the presumed stability envelope for a given stream would allow an engineer to design a more effective drainage project. If sinuosity is to be reduced, slopes should also be changed so that each new point (of the modified stream as plotted) remains within the stability envelope. In small agricultural watersheds, introduction of weirs or drop structures will allow straightening with less change in gradient.

Other basic tenets of fluvial processes and systems can also easily be applied to improvement projects to produce more efficient channelization. These ideas have existed for many years but have seldom been properly applied, possibly because of additional cost or possibly because of spatial limitations. However, other case studies and research have consistently shown that initially cheap construction methods are not necessarily the best nor ultimately the cheapest. It is safe to assume, therefore, that a greater initial investment will be justified by reduced long-term maintenance costs.

It is generally accepted that natural channels are maintained at bank-full discharge or flows near this stage. Leopold et al. (1964) believed that bank-full stage is attained once every 18 months, whereas others believe this ranges from 9 months to 2 years, depending on climate. Small watershed channelization projects, however, are usually designed to accommodate 25- to 50-year flood discharges. Thus the stream of bank-full or less discharge cannot be expected to maintain a channel designed for a 50-year flood. Yet the designed channel in many projects is trapezoidal in shape, with no specific design for a separate low-water channel for periods of less than flood flow. This practice may be the principal reason for low-water channel instability, increased midchannel bar deposition, and biologic sterility. Keller (1975) suggests the construction of a pilot channel, maintained by 2-year floods (or less), superimposed on a larger flood control channel, designed for the lower-frequency flood. If total flood control is desired, such larger channels would be analogous to construction of an artificial or confined floodplain. The low-water channel would be allowed to meander and would have nearly natural pool-riffle sequences to permit a productive aquatic habitat. The flood channel could even be vegetated to reduce wildlife displacement. In urban or suburban areas, if land is available, such artificial floodplains ideally could be used for recreation.

The Soil Conservation Service (1971) has proposed (and in some cases implemented) several techniques to improve fish productivity and conditions for other aquatic life by introducing a variety of flow conditions. Construction of artificial pool-riffle sequences is one method suggested; riffles are fabricated by placement of riprap at designated intervals along the stream course. Although the intent is sound, the induced riffles will be ineffective unless proper riffle spacing is determined for each stream. Furthermore, as generally designed the widened pool areas will soon silt up, as velocities are always reduced in such pools.

However, the application of the concept of convergent and divergent flow could be used to prevent siltation (Keller 1976). As discussed by Leliavsky (1966), diverging flow causes velocities to slow and thus material is deposited, whereas converging flow increases velocity and causes bed scour. In natural stream conditions, flow converges in pools and diverges in riffles, an idea quite inconsistent with high-velocity riffles and low velocities in pools at low flow. It appears, therefore, that these velocity conditions prevail only during low flow and undergo a velocity reversal in periods of high flow. That is, in high-flow conditions, relative velocities reverse, with higher velocities in pools that scour the bed. In diverging riffle sections at high flow, velocities are relatively lower, and deposition results. Construction of properly spaced pool-riffle sequences is desirable, but the use of convergent – divergent flow conditions should be applied to make such structures self-perpetuating.

Conditions observed on Big Pine Creek are not peculiar to this stream and can be compared to the few well-documented cases described in the literature. In 1910 a 33-mi (53-km)-long reach of Blackwater River in Johnson County, Missouri, was shortened by almost 18 mi (29 km). Stream gradients were thus increased from 8·8 ft/mi (4·25 m/km) to more than 16 ft/mi (7·7 m/km). Since channelization, scour and fill have been the dominant forms of readjustment with nearly 0·5 ft/yr (0·15 m/yr) of downcutting in the upper reaches, 3 ft/yr (0·9 m/yr) of lateral erosion, and excessive siltation downstream. The project has allowed more agricultural use of the floodplain in the river's upper reaches, but this benefit has been far outweighed by destruction of farmland and highway bridges downstream (Emerson 1971).

In 1961, a portion of the Peabody River in New Hampshire was relocated and shortened 850 ft (255 m). Channel gradients were changed from an original 52 ft/mi (25 m/km) to 80 ft/mi (39 m/km). Subsequently, in an attempt to restore the natural gradient, the stream alternately scoured and filled its bed, thus reducing the gradient to 75 ft/mi (36 m/km) after 2 years and 70 ft/mi (34 m/km) after 7 years. Considerable vertical and lateral scour accompanied the readjustment, with as much as 18 ft (5·4 m) of vertical scour within the reach immediately upstream of the channelized segment, and channel width increases as great as three to four times the original channel dimensions (Yearke 1971).

Between 1906 and 1920, Willow River in Iowa was modified to make adjacent lands fit for cultivation. In straightening, gradients were increased by almost 5 ft/mi (2·5 m/km) (more than 40%). Downstream siltation became prevalent, and additional maintenance was required in 1916 and 1941 because the channel bottom was silted above the level of tile drain outlets. The original objective of the project was achieved for the upper reaches of the Willow River, but not without secondary complications. Upstream flooding has not occurred since 1942, owing to entrenchment and widening of the channel. However, lower reaches have needed repeated remedial measures to retain necessary channel capacity. In this case it appears that insufficient consideration of and knowledge about the latent effects of channelization produced a degenerative and expensive drainage project (Daniels 1960).

The authors believe that in the cited cases, as well as on Big Pine Creek, damage could have been averted or reduced by proper planning and possibly a slightly greater initial investment. In almost every documented case, and certainly in numerous others that have gone unreported, nearly disastrous results have more than negated the positive aspects of stream channelization. We believe also that future channel modifications can be effective in increasing drainage capacity, but that certain cause and response effects inherent to fluvial systems must be recognized and accommodated in the design. If thresholds existing in the premodified stream are not exceeded, construction of a more stable, effective drainage project can be achieved, meanwhile retaining most of the original biotic, vegetational, and aesthetic qualities.

## REFERENCES

Benton County Court Records 1930–32. Ditch Docket 137.

Daniels, R.B. 1960. Entrenchment of the Willow Drainage Ditch, Harrison County, Iowa. *Am. Jour. Sci.* **258**, 161–76.

Davis, L.G. 1974. Floods in Indiana: technical manual for estimating their magnitude and frequency. *U.S. Geol. Survey Circ.* 710, 40 p.

Edgar, D.E. 1973. Geomorphic and hydraulic properties of laboratory rivers. M.S. thesis, Colorado State Univ., 156 p.

Emerson, J.W. 1971. Channelization: a case study. *Science* **173**, 325–6.

Gregory, K.J. and D.E. Walling 1973. *Drainage basin form and process: a geomorphological approach.* New York: John Wiley.

Holland, W.N. and G. Pickup 1976. Flume study of Knickpoint development in stratified sediment. *Geol. Soc. America Bull.* **87**, 76–82.

Leliavsky, S. 1966. *An introduction to fluvial hydraulics.* New York: Dover.

Leopold, L.B., M.G. Wolman and J.P. Miller 1964. *Fluvial processes in geomorphology.* San Francisco: W.H. Freeman.

Keller, E.A. 1972. Development of alluvial stream channels: a five stage model. *Geol. Soc. America Bull.* **83**, 1531-6.

Keller, E.A. 1975. Channelization: a search for a better way. *Geology* 3, 246-8.

Keller, E.A. 1976. Channelization: environmental, geomorphic, and engineering aspects. In *Geomorphology and engineering*, D.R. Coates (ed.), 115-40. Stroudsburg, Pa.: Dowden, Hutchinson & Ross.

Schumm, S.A. and H.R. Khan 1972. Experimental study of channel patterns. *Geol. Soc. America Bull.* **83**, 1755-70.

Soil Conservation Service 1971. *Planning and design of channel patterns.* Tech. Release 25, Chap. 7.

Yearke, L.W. 1971. River erosion due to channel relocation. *Civil Engineering* **41**, 39-40.

# 16

# EVALUATING AQUATIC HABITAT
# USING STREAM NETWORK STRUCTURE
# AND STREAMFLOW PREDICTIONS

*John B. Stall and Edwin E. Herricks*

### ABSTRACT

A one-year field study has been made of the 8301-km² basin of the Little Wabash River in Illinois to determine instream flow needs for aquatic life. The stream system structure follows Horton's laws fairly well. Three reaches of about 1·5 km each were selected for detailed study of hydraulics, hydrology, and aquatic life. These reaches were selected to be representative of the entire stream system. The analytical tool used at each reach was the IFG Incremental Methodology developed by the Instream Flow Group of the U.S. Fish and Wildlife Service. Detailed measurements of depth, velocity, and bed material were taken at each reach. The methodology relates the area of fish habitat calculated from these inputs based on preference curves of various fish species and life stages. Hydraulic geometry equations are given to allow calculation of stream discharge, cross-sectional area, width, depth, and velocity at any location in the stream system. Results show the amount of available habitat as related to streamflow and as the amount of habitat varies throughout the various months of the year. The results are considered meaningful and useful for managing streamflow for instream uses.

## INTRODUCTION

### Need

An understanding of stream network structure, hydraulics, and hydrology can contribute to understanding the nature and maintenance of aquatic ecosystems. This allows an improvement in the use and maintenance of *instream flows*; it allows better development and use of our total water resources. The Federal Water Resources Planning Act of 1965 created the Water Resources Council to establish principles of water policy and to prepare an assessment of water availability. The first national assessment was published in 1968 and the second assessment in 1977. The Council early recognized the lack of methodology to determine the effects of streamflow on biological systems. The second assessment identified several conflicts between water uses *in* the stream and water uses *outside* the stream. This conflict is greatest in western United States, where annual runoff is less than 0·25 m (10 in.).

## Assessment procedures

In making the national water assessments, water use and water supply were evaluated in detail for about 100 Aggregated Subareas of the nation. Eisel (1979) described the crude estimates made in the second assessment, of instream flow needs for fish. The Montana method was used. It and other methods have been described by Stalnaker and Arnette (1976). The Montana method uses these principles: (a) if the streamflow is maintained at 60% of the annual mean flow it will provide outstanding habitat for aquatic life; (b) a streamflow of 30% of the annual mean will provide good habitat for most aquatic life; and (c) streamflow at 10% of the annual mean can provide only short-term survival habitat for most aquatic life. The method was recognized as needing much improvement.

The most advanced methodology available in 1980 to relate instream flows to aquatic habitat which has replaced the Montana method in many instances is that developed by the Instream Flow Group, the IFG, of the Office of Biological Sciences of the U.S. Fish and Wildlife Service at Ft. Collins, Colorado. The system is termed the IFG Incremental Methodology and utilizes an elaborate computer program called PHABSIM. It has been described by Stalnaker (1980); and an overall critique of the method by fifty selected scientists is presented by Smith (1979). This IFG Incremental Methodology was used on each of the three reaches studied on the Little Wabash River.

The *Incremental Methodology* develops a display of the various *increments* of fisheries habitat provided in a stream reach for various discharges. It is one of several impact analysis procedures recently developed by the U.S. Fish and Wildlife Service. It depends on an accurate description of available physical conditions in the stream; it then translates that information into available habitat for fish and wildlife. With the Incremental Methodology a hydraulic simulation provides depth and velocity information for a cross section with known substrate conditions. The cross-sectional data are extrapolated, producing areal estimates of depth, velocity, and substrate. Extrapolations are carried half the distance to the adjoining cross section. This defines a set of habitat conditions which are then combined with fish preference for those conditions. The linkage between the hydraulic simulation and habitat suitability is an independently derived species preference curve.

A typical curve is illustrated in Figure 1. Because all organisms are subject to a range of environmental conditions, testing or observation can define the optimum level for a parameter as well as tolerance ranges. As Figure 1 illustrates, the organism's optima for velocity is between 0·0 and 0·07 m/s. As velocity increases, conditions become less than optimum and the organism will show less preference for that velocity. At some point a threshold is reached that will usually eliminate the organism from the habitat. Thus each organism has a range of preference (or tolerance) for single environmental parameters. Through experimental development of preference curves a link can be made between hydraulic variables and habitat potential.

Because stream ecosystems are complex, preference curves must be developed for many species. Even for a single species, analysis must include the habitat preferences of each life stage. Newly hatched fry, juveniles, and adults may all have slightly different habitat requirements or preferences. For example, Figure 2 shows preference curves for the bluegill sunfish. Spawning will generally occur at a depth of 1 to 1·5 m, at velocities less than 0·2 m/s, and over a substrate of mud or sand. To be incorporated into the Incremental Methodology, habitat requirements must be determined for each life stage. For example, bluegill spawn in late April through May with fry developing quickly to

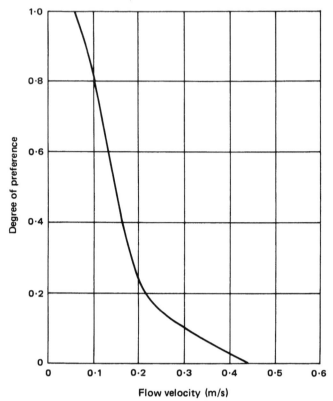

**Figure 1**  Fish preference curve.

juveniles, with adults present throughout the remainder of the year.

In the Incremental Methodology preference values for any life stage are assumed to be independent and multiplied together. This product of preference values is used to modify the areal estimate made from the hydraulic simulation to produce the actual square meters of usable habitat available in this reach of stream.

## The basin

The Illinois Division of Water Resources has overall responsibility for planning, and in some cases regulating, water use in Illinois. They desired to have the instream flow needs for fish determined by a method that would provide maximum technical support.

The basin selected for a detailed 1-year study was the Little Wabash River in southeast Illinois, which is typical of Illinois conditions. It begins near Mattoon and flows about 385 km southward; it empties into the Wabash River and has a total basin area of 8301 km². Figure 3 shows the bed profile. Three 1·5-km reaches of the river were selected for aquatic and hydraulic observations. These reaches are titled for reference according to the nearest city: Effingham, Clay City, and Golden Gate. Figure 3 shows that the bed slopes of these reaches are representative of the river profile.

The Horton–Strahler parameters for the stream network structure for the Little Wabash River stream system were published earlier by Stall and Fok (1968) and are

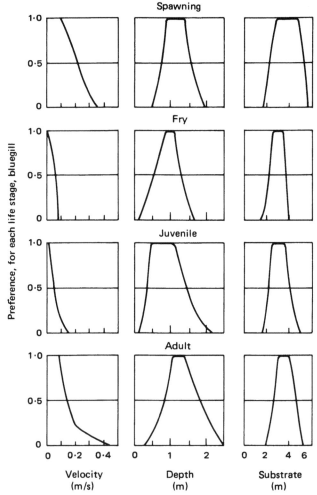

**Figure 2** Preference curves for bluegill for velocity, depth, and substrate in a stream.

shown graphically in Figure 4. The number of streams, length, and slope seem to follow Horton's laws to a fair degree. This gave us some confidence that representative reaches to be studied could be selected and later used to generalize stream conditions throughout the entire stream system. This confidence is based on the fact that the stream network does follow a consistent pattern of relationships among number, length, and slope of stream segments.

The Effingham Reach is on a third-order stream; the Clay City Reach is on a fifth-order stream, and the Golden Gate Reach is on a sixth-order stream. The physical nature of the stream, such as its width, depth, and slope, can be interpreted with confidence for other stream orders not sampled, because of the consistency of the variation of these parameters as revealed by the consistency of Horton's laws. The five hydraulic geometry equations representing the interrelationships of these physical factors are given later in this paper.

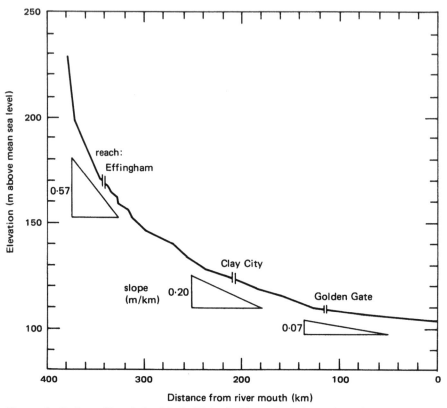

**Figure 3**  Bed profile of the Little Wabash River. Bed slopes in m/km are shown as triangles for the three reaches of the river studied in this project.

## STREAM NETWORK AND STREAMFLOW PREDICTIONS

The primary method selected to determine instream flow needs of the Little Wabash River was the Incremental Methodology. Application of the Incremental Methodology requires measurement of depth, velocity, the particle size of the substrate, and water surface evaluation at cross sections in the representative reach but can be supplemented by additional basin hydraulic analysis such as the development of hydraulic rating curves described below and actual biological sampling. Representative reaches were selected using both land and aerial reconnaissance. Final placement of cross sections was made based on bed profiles obtained by use of a recording depth sounder and boat. Cross sections were placed based on two criteria. The first was identification of bed features which controlled channel hydraulics. The second was description of fisheries habitat. Thus cross sections were not evenly spaced, but placed to describe channel characteristics typical of a longer river reach.

### Hydraulics

The IFG Incremental Methodology is based on developing a good understanding of the hydraulic features of the reach being studied. This is done by taking field measurements

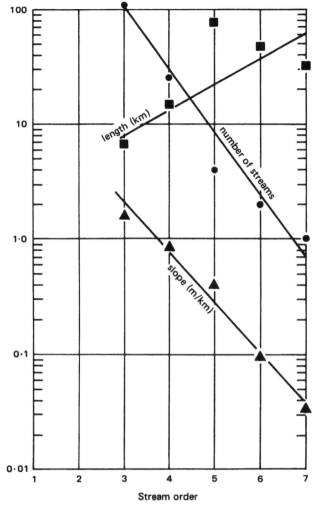

**Figure 4** Horton relationships for the Little Wabash
River stream system (data from Stall and Fok, 1968).

of water depth and velocity, calculating hydraulic roughness, and calculating a water
surface profile through the reach. The procedures for the stream hydraulic work are well
described in how-to-do-it fashion by Bovee and Milhous (1978). An authoritative description of detailed observations of hydraulics of a natural river in Illinois is provided by
Bhowmik (1979); this was of much value to the authors.

At each of the three reaches a set of ten cross sections were studied at various intervals
along the 1·5-km reach to represent various hydraulic and habitat features. The ends of
these cross sections were marked with permanent concrete benchmarks. The surface of
the stream was mapped. All hydraulic and aquatic measurements were taken along these
cross sections. Each reach was visited three times to measure discharge at conditions as
nearly as possible at (1) high flow, or bankfull; (2) medium flow; and (3) low flow, or
pool-and-riffle condition. Discharge was measured with a current meter at about fifteen

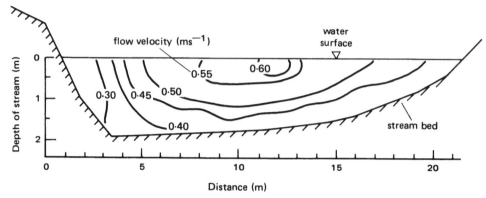

**Figure 5** Isovels, or lines of equal flow velocity, in the Clay City reach, measured at a discharge of 14·9 m³/s.

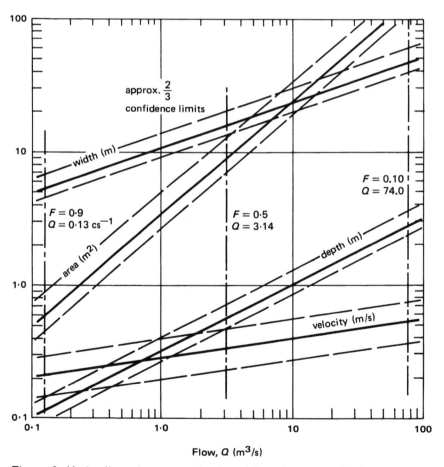

**Figure 6** Hydraulic rating curves developed from long records of streamflow measurement on the Little Wabash River near Clay City.

to twenty-five vertical sections about 1 m apart across the river. This technique is used extensively by the U.S. Geological Survey (USGS) throughout the nation at stream gauging stations and is described by Buchanan and Somers (1973). Figure 5 illustrates a typical cross section in the Clay City reach at a rather high discharge. Shown are *isovels*, lines of equal flow velocity. There is considerable variability in flow velocity at various locations in the cross section. These flow velocities, as has been described, affect fish habitat, and are a critical input to the PHABSIM program. In the field, one sample was taken of bed material on each cross section and the size distribution was determined by sieve-hydrometer techniques. This provided input to the PHABSIM program regarding the substrate of the stream.

A long-term stream gauging record is available for the Little Wabash River below Clay City, within the Clay City reach. Since 1915 the USGS has taken several hundred discharge measurements and Figure 6 shows hydraulic rating curves for this gauge. The horizontal scale is discharge, $Q$. The solid lines in the graph show how the stream's cross-sectional area $A$, width $W$, depth $D$, and flow velocity $V$ increase as the flow increases. The dashed vertical line represents a discharge of $0.13$ m³/s. It has a flow frequency $F$ of $0.90$, meaning that on 90% of the days each year the discharge exceeds $0.13$ m³/s. The solid lines intersect this dashed line at the expected value, of area, width, depth, and velocity of the stream when $Q = 0.13$ m³/s. Other lines show conditions for median flow, $F = 0.50$; and for $F = 0.10$, which is high or *bankfull* flow. Results such as these can only be developed from records at a long-term stream gauging station. They are required to develop the hydraulic geometry relations for the basis, described next.

## Hydraulic geometry

The IFG Incremental Methodology, accomplished by the computer program PHABSIM, determines the aquatic habitat available, based on the depth, flow velocity, and substrate of the stream in the reach studied. To extend these results to the entire 8301-km² basin of the Little Wabash River, use was made of the stream network structure determined by the hydraulic geometry of the stream system.

Stall and Fok (1968) described a consistent pattern for all eighteen river basins in Illinois, in which the width, depth, and flow velocity change along the course of a stream with a constant frequency of discharge. This was shown for ten river basins of the United States by Stall and Yang (1972). These channel characteristics are termed *hydraulic geometry* and constitute an interdependent system that is described by a series of graphs having simple form, or by equations. Stall (1980) has described how to derive hydraulic geometry equations for a river basin. In Illinois, data from 166 stream gauging stations were used by Stall and Fok (1968). In the Little Wabash River basin, four gauges were used.

The hydraulic geometry equations for the Little Wabash River system are given as follows:

$\ln Q = -2.57 - 7.90F + 0.96 \ln A_d$
$\ln A = \phantom{-}0.28 - 6.72F + 0.66 \ln A_d$
$\ln V = -2.86 - 1.18F + 0.30 \ln A_d$
$\ln W = \phantom{-}1.36 - 2.68F + 0.34 \ln A_d$
$\ln D = -1.09 - 4.04F + 0.32 \ln A_d$

where $Q$ = flow, m³/s
$\quad$ $A$ = stream cross-sectional area, m²
$\quad$ $V$ = flow velocity, m/s
$\quad$ $W$ = stream width, m
$\quad$ $D$ = average depth of the stream,
$\quad$ $F$ = flow frequency, percent of days per year that flow value is exceeded, expressed as a decimal; $F = 0.10$ is the flow that is exceeded only 10% of the days in a year
$\quad$ $A_d$ = drainage area, km²

and the designation ln signifies logarithm to the base e. In these equations the range of flow frequency is limited such that $0.10 < F < 0.90$.

Using these equations it is possible to select a flow frequency value such as $F = 0.10$ (about *bankfull* flow) to determine the drainage area $A_d$ at a point on any stream in the basin; then the equations can be used to calculate representative values of the other variables at that point on the stream. Figure 7 shows how stream depth is related to flow frequency $F$ and to the drainage area $A_d$ throughout the Little Wabash River basin.

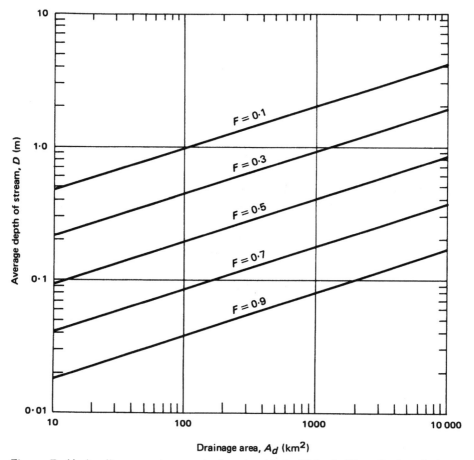

**Figure 7** Hydraulic geometry graph for the Little Wabash River Basin relating stream depth, $D$, to flow frequency, $F$, and drainage area, $A_d$.

## Hydrology

The seasonal variation in streamflow affects the aquatic habitat. For the Clay City reach an analysis has been made of this variation using the 63 years of streamflow data available. The results are shown in Figure 8. The horizontal scale is the months of the year; the vertical scale is flow. For every month the median flow for that month is shown as the dashed line; the January flow shown is exceeded in 50% of all Januaries; the solid lines show the flows exceeded in 1% of all Januaries, in 99% of all Januaries, and so forth. The three horizontal dashed lines represent, for the entire year, the flows calculated by the hydraulic geometry equations for high flow, $F = 0 \cdot 10$; for the annual median flow, $F = 0 \cdot 50$; and for low flow, $F = 0 \cdot 90$, which is a pool-and-riffle condition.

The development and maintenance of the existing aquatic life in the Clay City reach has occurred under the pattern of seasonal flow variability depicted in Figure 8. The continued maintenance of this aquatic life can be accomplished best by our understanding of these limits of streamflow variability and how this affects aquatic habitat.

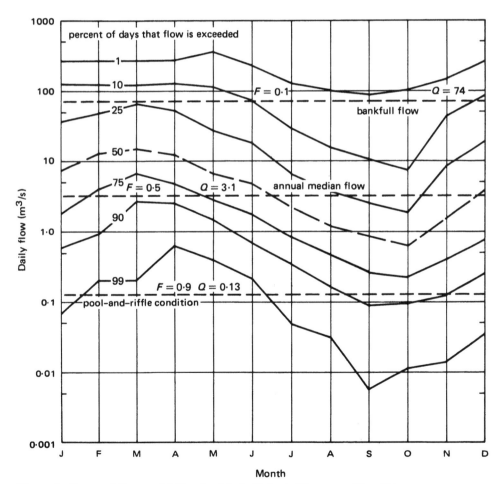

**Figure 8**  Seasonal flow variability for Little Wabash River near Clay City.

## Aquatic life

Although fish preference curves similar to those in Figure 2 are contained in the computer package and can be used in analysis procedures, additional biological analysis was made. To determine the composition of stream fisheries, each reach was sampled on one or more occasions by electrofishing. In addition, detailed sampling was conducted to develop preference curves for indigenous fishes. The process of curve development is complicated and conducted independently of Incremental Methodology data collection. Typical procedures for preference curve development include random selection of a stream reach and habitats within that reach. Probes that deliver an electric current were placed in the water and an electric current supplied. Fish in the electric field were stunned and rose to the surface. Fish were identified (using Smith 1979), weighed and measured, and the site of their appearance marked. Measurements of depth, velocity, temperature, and substrate were made for each fish. From these data preference curves were constructed.

## HABITAT EVALUATION RESULTS

For the Clay City reach, hydraulic measurements, field observations of existing aquatic life and fish preference curves from reference sources were used as input for the PHABSIM program. The output results are shown in Figure 9 for the Bluegill species, for the four life stages of adult, spawning, juvenile, and fry. The available habitat in square meters per 1000 m of stream length is shown as a function of streamflow. At low discharge there is a relatively high amount of available habitat. As discharge increases, the available habitat decreases; this decrease is much greater for the fry than for the adult.

The data in Figure 9 were used to develop the seasonal change in available habitat; this is shown in Figure 10 for the bluegill. Here it is seen that all year long there seems to be plenty of available habitat for the adult. There is also plenty of available habitat for spawning during the usual April through June season. The fry would have a much harder time surviving the season because Figure 10 shows a small amount of available habitat during the months (April through June) that fry exist. The available habitat seems also to be somewhat limiting to the juvenile life stage.

Using only bluegill predictions, it is possible to recommend flows that would support adults as well as providing habitat for juveniles and fry. In the Little Wabash, actual fisheries analysis indicated low populations of bluegill and higher populations of other fish. Sound management would dictate that all important species be evaluated and decisions made only after a full range of discharge options were considered.

## CONCLUSIONS

The IFG Incremental Methodology provides meaningful results depicting available aquatic habitat. It provides an understanding of hydraulic, hydrologic, and biological factors that can aid in making management decisions for instream flows. Trade-offs can be evaluated. A flow variability regime can be developed which will best satisfy the instream flow needs of fish, recreation, water quality, and estuarine inflows.

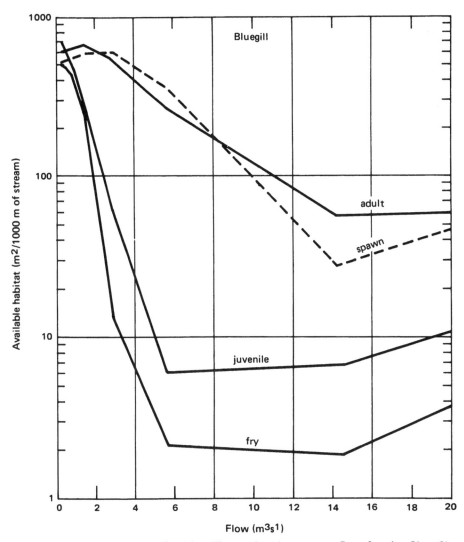

**Figure 9** Aquatic habitat for bluegill as related to streamflow for the Clay City reach.

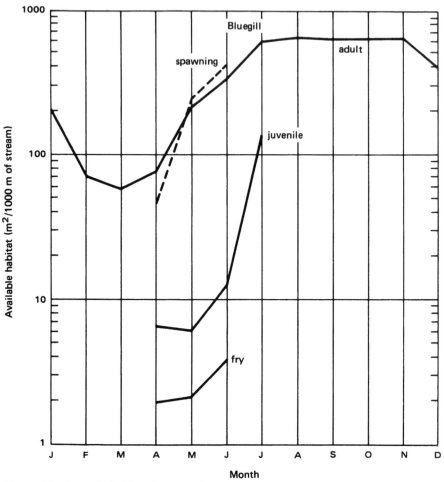

**Figure 10**   Aquatic habitat for bluegill as it varies throughout the year at the Clay City reach.

## REFERENCES

Bhowmik, N.G. 1979. *Hydraulics of flow in the Kaskaskia River, Illinois.* Illinois State Water Survey Rept. Invest. 91, Urbana, 116 p.

Bovee, K.D. and R. Milhous 1978. *Hydraulic simulation in instream flow studies: theory and techniques.* Inf. Paper Cooperative Instream Flow Group No. 5, Ft. Collins, Colo., 131 p.

Buchanan, T.J. and W.P. Somers 1973. *Discharge measurements at gauging stations.* U.S. Geol. Survey Techniques in Water Resources Invest., Bk. 3, Chap. A8, 65 pp., Washington, D.C.

Eisel, L.M. 1979. Instream flow needs and water resources planning. In *Proceedings, workshop in instream flow habitat criteria and modeling*, G.L. Smith (ed.), 4–15. Colorado Water Resources Research Inst. Inf. Ser. 40, Ft. Collins, Colo.

Smith, G.L. 1979. *Proceedings, workshop in instream flow habitat criteria and modeling.* Colorado Water Resources Research Inst. Inf. Ser. 40, Ft. Collins, Colo. 244 p.

Smith, P.W. 1979. *The fishes of Illinois.* Urbana, Ill.: University of Illinois Press, 314 p.

Stall, J.B. 1980. A model structure for stream systems. In *Computer and physical modeling in hydraulic engineering,* G. Ashton (ed.), 98-106. New York: American Society of Civil Engineers.

Stall, J.B. and Y.-S. Fok 1968. *Hydraulic geometry of Illinois streams.* Illinois State Water Survey Contract Rept. 92, Urbana, 47 p.

Stall, J.B. and C.T. Yang 1972. *Hydraulic geometry and low stream-flow regimen.* Illinois State Water Survey Contract Rept. 131, Urbana, 73 p.

Stalnaker, C.B. 1980. The use of habitat structure preferenda for establishing flow regulation necessary for maintenance of fish habitat. In *Ecology of regulated streams,* J.V. Ward and J.A. Stanford (eds.), 321-37. New York: Plenum Press.

Stalnaker, C.B. and J.L. Arnette 1976. *Methodologies for the determination of stream resource flow requirements: an assessment.* Western Water Allocation, Office of Biological Services, U.S. Fish and Wildlife Service, 199 p.

Milton Keynes UK
Ingram Content Group UK Ltd.
UKHW040107071024
449327UK00019B/880